HERE BE DRAGONS

HERE BE
DRAGONS

How the study of animal and plant distributions
revolutionized our views of life and Earth

DENNIS McCARTHY

OXFORD
UNIVERSITY PRESS

OXFORD
UNIVERSITY PRESS

Great Clarendon Street, Oxford OX2 6DP

Oxford University Press is a department of the University of Oxford.
It furthers the University's objective of excellence in research, scholarship,
and education by publishing worldwide in

Oxford New York

Auckland Cape Town Dar es Salaam Hong Kong Karachi
Kuala Lumpur Madrid Melbourne Mexico City Nairobi
New Delhi Shanghai Taipei Toronto

With offices in

Argentina Austria Brazil Chile Czech Republic France Greece
Guatemala Hungary Italy Japan Poland Portugal Singapore
South Korea Switzerland Thailand Turkey Ukraine Vietnam

Oxford is a registered trade mark of Oxford University Press
in the UK and in certain other countries

Published in the United States
by Oxford University Press Inc., New York

British Library Cataloguing in Publication Data

Data available

Library of Congress Cataloging in Publication Data

Data available

McCarthy, Dennis, 1964-
Here be dragons : how the study of animal and plant distributions revolutionized
our views of life and earth / Dennis McCarthy
  p. cm.
Includes bibliographical references.
ISBN 978–0–19–954246–8
1. Biogeography. 2. Plate tectonics. 3. Evolution (Biology) I. Title.
QH84.M34 2009
578.09–dc22

Typeset by SPI Publisher Services, Pondicherry, India
Printed in Great Britain
on acid-free paper by
Clays Ltd., St Ives Plc

ISBN 978–0–19–954246–8

1 3 5 7 9 10 8 6 4 2

To my wife,
Lori,
who has lovingly and tirelessly
supported me in all my ambitions

ACKNOWLEDGMENTS

I thank Malte Ebach, who pushed me forward toward publication. The selflessness of his help continues to astound. I also thank Latha Menon, whose kindness and amiability belie the intimidating scope of her expertise and the soundness of her editorial diligence. She has saved me from numerous errors—factual, logical, grammatical—and the changes she has suggested have greatly improved the book. I also appreciate Mara Huber for her generous provision of research materials and Alan Templeton and Gonzalo Giribet for perusing and correcting those passages that pertain to their work. I also extend gratitude to Elliot Lim and Sergio A. Marenssi, who provided such beautiful figures.

On a personal level, I would never have had the time to work on this book without constant help and understanding from my family—and especially the numerous sacrifices of my wife, Lori, and my mother, Gloria. And I send a special thank you to Nicole, Meagan, Kennedy, and Griffin for adding such sweetness, fun, and delight to my life.

CONTENTS

CONTENTS

LIST OF ILLUSTRATIONS

LIST OF ILLUSTRATIONS

LIST OF ILLUSTRATIONS

The three crustal age figures of Color Plates 1, 2, and 3 have been provided by the image author: Elliot Lim, Cooperative Institute for Research in Environmental Sciences (CIRES), NOAA National Geophysical Data Center (NGDC), Marine Geology and Geophysics Division. Data source: R.D. Muller *et al.*, "Age, spreading rates and spreading symmetry of the world's ocean crust," *Geochemistry, Geophysics, Geosystems*, 9 (2008), Q04006, doi:10.1029/2007GC001743

The figure in Color Plate 4, *The giant ground sloth, Megatherium,* was reproduced with permission from the Natural History Museum of London.

PREFACE: "THAT GRAND SUBJECT"

Biogeography as the fountainhead of scientific discovery and the unifying theory of life and Earth

Within the first few minutes of thumbing through the book *Foundations of Biogeography: Classic Papers with Commentaries*, I came across one of the most astonishing facts on the development of modern science that I had ever encountered. The book is a squat, unwieldy collection of papers from a variety of specialists all trying to explain why certain plants and animals are found in some places and not others, all trying to highlight and explicate patterns in the distribution of life. But the most remarkable thing about *Foundations of Biogeography*, the discovery I found so shocking, appeared not in any of its papers but in its Table of Contents. Although the book focuses exclusively on contributions to the little known subject of biogeography, its list of authors reads like a *Who's Who* of scientific revolutions. The following are a few of the scientists who had penned some of the collected classics:

Carolus Linnaeus, the father of modern taxonomy
Charles Darwin, co-discoverer of the theory of evolution
Alfred Russel Wallace, co-discoverer of the theory of evolution
Alfred Wegener, the father of continental drift

PREFACE: "THAT GRAND SUBJECT"

E. O. Wilson, the modern father of sociobiology and
 author of *Consilience*
Jared M. Diamond, author of *Guns, Germs, and Steel*

Incredibly, before these researchers had all helped raze conventional assumptions in biology, geology, sociology, and anthropology, they had all developed an expertise in the geographical distributions of flora and fauna. Certainly, no other science boasts such an illustrious pedigree. Other scientific compendia, like *The World of the Atom, Foundations of Animal Behavior,* and *History of Mechanics,* do include papers from a number of prodigies, but the renown of these scientists almost always came from their work in that same discipline—physicists in physics, chemists in chemistry, etc. Only in biogeography do we find so many different geniuses who have revolutionized *other* fields of science. Try to imagine six researchers who have penned masterpieces in some intellectual discipline—say, linguistics or statistical mechanics or psychology—branching out and overthrowing the foundational tenets of four other disciplines, and you may get an idea of how special biogeography is.

One of the main purposes of this book is to answer why this is—to explain why so many revolutionaries in so many disparate fields of thought were all specialists in the little known field of biogeography. Charles Darwin likely became one of the first to note the transcendental significance of biogeography when, in an 1845 letter to J. D. Hooker, he lauded it as "that grand subject, that almost keystone of the laws of creation." Fourteen years later, Darwin would publish *On the Origin of Species* with two chapters devoted to "Geographical Distribution." In the early twentieth century, biogeography would return to the forefront of another major scientific revolution, again occupying a chapter in one of the most important books in the history of science—Alfred Wegener's *The Origin of Continents and Oceans*. In 1997, Jared Diamond would challenge orthodox views of anthropology by putting forth a novel explanation for the differences in the development of human societies on different continents—a view Diamond founded on the principles of biogeography. As we shall see, it is likely that, excepting the principle of material causality, no other known tenet or group of facts has proved more fruitful to the intellectual progress of the human race than the distributional patterns of plants and animals.

Chapter 1, "Galápagan Epiphany," describes the voyages, both global and intellectual, of Charles Darwin; Chapter 2, "The *Mesosaurus* Problem," explores the journeys of Alfred Russel Wallace, Alfred Wegener, and Alexander du Toit. The chapters will detail how the biogeographical evidence that they discovered led them to their revolutionary viewpoints. Why are frogs and newts so often absent from oceanic islands—and why was that so important to the mystery of the origin of species? Why are fossils of the long-snouted aquatic reptile *Mesosaurus* exclusively found in South America and South Africa—and what does that have to do with planetary science? As the development of the theories of evolution and continental drift confirm, these researchers were not simply biogeographers by hobby; they repeatedly used the implications of distributional facts to govern their earth-changing conclusions. These chapters will explain why the deceptively simple question "How did plants and animals end up where they are?" has been one of the most important in the history of science.

The next four chapters will use the history of a number of exotic creatures to show how the theories of evolution and plate tectonics merge in the meta-theory of biogeography. As Darwin,

Wallace, Wegener, and du Toit all understood: life and Earth evolve together. And plants and animals are as much a part of the landscape as the mountains or rivers or plains that they inhabit. With unwavering fidelity, organisms will reflect the various changes, whether geological or climatic, that occur throughout a region. Plants and animals, as we shall see, are the discreet and faithful keepers of Earth history.

While many scientific fields have become a conglomeration of specializations, ever narrowing the focus of the researcher's gaze, the modern biogeographer continues to zoom outward, often illuminating the broader patterns and principles that occur on a continental, oceanic, or even global scale. In brief, modern biogeography provides a panoramic portrait of evolution and global processes working in concert. Continents rift along volcanic ridges; oceans flow into the gaps; islands spew into existence from volcanoes of the deep; and mountains thrust upward as continental regions converge and plates begin to fold. These earthly upheavals lead to changes in climates: sea levels rise or fall; glaciers advance or retreat; deserts become fertile; or rainforests turn barren. The evolving landscapes and new barriers divide various species of plants and animals, insects and fish, isolating certain groups in new environments, pushing them into different climes, mixing them with new predators and food sources, creating new hazards and eliminating old ones. Organisms must continuously adapt or perish on this always dynamic and often violent Earth.

Some of these chapters will include illustrations and maps that highlight the relationships between the animals and the landscape, between the green and the granite, between evolutionary changes and geological events. This broader perspective will underscore why biogeography is not simply the place where evolution, plate tectonics, oceanography, and climatology meet; it is the theoretical framework that subsumes all these theories. While physicists are still reaching for that grand unified theory that will unite gravity (general relativity) with electromagnetism (quantum mechanics), biogeography already provides that grand unified theory of life (evolution) and Earth (plate tectonics and the geosciences).

Chapter 3, "Pygmy Mammoths and Mysterious Islands," will explore the relationships between a variety of unique organisms and the remote islands they inhabit, including the evolution of a dwarf species of mammoth on the Channel Islands just outside of California. The chapter will also help answer the questions "Why Galápagos?" and "Why finches?" While many popular science authors have discussed Darwin's trip to these South American Islands, few have tried to convey the biogeographical causes that had turned the island group into such a remarkable Darwinian laboratory. The chapter will also discuss new discoveries that finally expose the secret behind the evolutionary explosion of Darwin's finches. As we shall see, Darwin's visit to the Galápagos group was the last link in a "perfect storm" of events— a fortuitous convergence of the perfect scientist, the perfect islands, and the perfect birds, all coming together to help spark what some consider "the greatest idea ever to occur to a human mind."[1]

Chapter 4, "The Volcanic Ring That Changed the World," explores how the formation of a system of cracks in the seafloor around Antarctica has resulted in the isolation of the Southern Hemisphere landmasses, producing many of the major organic patterns observed today. The consequences of the resulting continental arrangement on plants, animals, and even human civilization would be difficult to overestimate—and will be discussed throughout the course of the book. Why is the Southern Hemisphere so oceanic? Why have northern mammals managed to dominate the world? What is the reason for the technological gap between Eurasians and the native societies of Australia, South America, and New Zealand? The answers, as we shall see, all lead back to the ring of volcanic ridges that currently surrounds the southernmost continent.

Chapter 5, "The Bloody Fall of South America and the Last of the Triassic Beak-Headed Reptiles," describes the isolating consequences of the breakup of Gondwana that helped occasion pockets of stability and stasis. While many of the processes that have steadily sculpted our planet have served as the engine for speciation, still other geological events have supported the long-lasting preservation of certain plants, reptiles, mammals, and amphibians. As will be shown, the long isolation of New Zealand Australia, and South America has provided sanctuary for a number of "living fossils," protecting them from the more recent and often fiercer or sturdier by-products of evolution from North America, Eurasia, and Africa. But occasionally, when those barriers are removed, as when the Isthmus of Panama emerged from the sea and connected the Americas, the consequences for the fauna of the formally isolated region are often tragic.

Chapter 6, "Enchanted Waters," will explore the biogeographical patterns within the seas, showing that the same distributional principles that govern life above sea level also apply to the creatures living below. The chapter will also detail evolution operating in secluded places like Lake Baikal, that shimmering jewel of Siberia that has often been referred to as the "Galápagos of Russia." Such isolated lakes are the aquatic analogue of oceanic islands, often separating their flora and fauna from the rest of the world and leading to many new and seemingly alien organic forms.

The controversial subject of the biogeography of the human race will be the subject of Chapter 7, "The Battle Over Eden," with a focus on the theoretical storm over the location of human origins, the recent discovery of the "hobbits" of Flores Island, and Jared Diamond's new path-forging view of the development of civilizations from *Guns, Germs, and Steel*. As Diamond has shown, biogeography has been as influential in determining the fates of human societies as it has in shaping our scientific views. And once again we discover the importance of continental placement in shaping life history.

"Here Be Dragons" (or, at times, "Here There Be Dragons") is a semantic legend, an alleged cartographical slogan that many people incorrectly believe was scrawled on the unfamiliar territories of antique maps. In reality, the phrase is unknown from ancient charts, although you will find many similar biogeographical pronouncements like "in these places scorpions are born" or "here lions abound." The only known place where historians have been able to locate the phrase "Here Be Dragons" is on the Hunt-Lenox Globe (*c*.1506), a small copper sphere, a little less than 5 inches (13 cm) in diameter, now on display at the New York Public Library. If you look closely at the Eastern Hemisphere of this softball-sized, metallic globe—one of the first to be constructed after Columbus's journey to the New World—you will find the words HC SVNT DRACONES (*Hic Sunt Dracones* or "Here Be Dragons") stamped upon Southeast Asia. Many people had believed the phrase was a warning to those journeying to strange lands, but given the propensity of medieval map-makers to attempt to locate other compelling creatures—including elephants, polar bears, scorpions, lions—"Here Be Dragons" was probably another archaic effort to chart an organic distribution. The Komodo and Flores Islands of Southeast Asia are home to the giant, flesh-eating Komodo dragons, the largest of all living reptiles, and stories of its ferocity may have led to the famous detail on the Hunt-Lenox Globe. Thus, "Here Be Dragons" is probably one of the most well-known biogeographical comments in history—and the first one to be placed on a post-Columbian globe.

Chapter 8 summarizes the past, present, and future of biogeography, from the very first distributional chartings like *Hic Sunt Dracones* to the new tools and analytical devices now available to biogeographers that have led to recent and exciting discoveries. This final chapter will also use the facts and principles detailed in the book to show how biogeography relates to our everyday life and current environment. All the elements of the organic world surrounding you now are part of larger distributional patterns that have been shaped by Earthly forces, so understanding biogeography will help you find the history of your region in the plants and animals that you see daily. For example, a biogeographical investigation of the traditional fare at a Hawaiian luau—the pig, bananas, coconuts, chicken, etc.— illuminates the extraordinary history of Polynesian society, helping confirm once and for all the extraordinary extent of their Pacific conquests. The chapter will also summarize how and why a small group of experts in the distributions of plants and animals has managed to answer so many of our fundamental questions about the origins of life and species, the origins of continents and oceans, the origins of civilizations and technology. Certainly, all studies of natural phenomena are significant and have added immeasurably to our current views of life and Earth. But biogeography stands apart. The wide-ranging influence of plant and animal distributions in shaping modern viewpoints simply has no parallel in science and has more than justified Darwin's faith in its importance. Biogeography truly has been "the keystone of creation."

CHAPTER 1

Galápagan Epiphany

How the distributions of plants and animals led Darwin to the theory of evolution

"I did not then in the least doubt the strict and literal truth of every word in the Bible."
    —Charles Darwin, from his autobiography,[1] on the years
        1828–31 just preceding his voyage around the world

"When on board H. M. S. 'Beagle,' as naturalist, I was much struck with certain facts in the distribution of the inhabitants of South America, and in the geological relations of the present to the past inhabitants of that continent. These facts seemed to me to throw some light on the origin of species."
    —The first sentences of Charles Darwin's
        On the Origin of Species (1859)

W hen Charles Darwin first boarded the HMS *Beagle*, eager to embark on the most colossal adventure of his life, he was still a conventionally religious young man. Like most naturalists of his time, Darwin had no doubt that God was the Creator of the Universe, the Grand Designer of both the celestial and the earthly. He knew that God created the stars and that God conjured the organic, breathing life into species in all their forms and placing them in the various regions of the Earth. Darwin believed that God wisely chose to set camels in the deserts, mountain

1

goats on mountains, whales in the ocean, and sheep in grassy fields—and that God bestowed upon each plant or animal, fish or insect, all the peculiar and helpful accoutrements that each required to flourish in its particular habitat. This apparent evidence of the mind of God in every aspect of the organic world, a view detailed in William Paley's influential book, *Natural Theology*, enabled one to study God's awesome plan by carefully examining the creatures that were His handiwork. This became the predominant goal of most natural historians of the time. And this was Darwin's goal.[2]

But as Darwin journeyed with the *Beagle* around the world, faithfully collecting and recording the flora and fauna from each exotic stopover, he began to notice a problem. Yes, natural theology adequately explained why so many different organisms had traits that were peculiarly adapted to their specific location and lifestyle. The long, thin beaks of hummingbirds, the webbed feet of beavers, the sticky silk webs of spiders—all were evidence of God's careful and inventive arranging. What natural theology could not explain, however, and what Darwin did not expect to discover, was a persistent global pattern in the distribution of life, a continuous organic thread running from region to region, stretching uninterrupted through space and time. What natural theology could not explain, Darwin had started to realize, was biogeography.

Darwin knew that the basic premises of independent creation permit, or even demand, the occurrence of identical or closely related species in similar but widely separated environments. We should, in the creation-based view of the pre-Victorian naturalist, find matching species on opposite sides of the world as long as the environments are sufficiently alike—similar plants and animals, for example, in the Arabian Desert and the Mojave

Desert of North America or in the icy worlds of the Arctic and Antarctica. But this contrasted with the organic world that Darwin had begun to explore. Everywhere around the globe, taxa that were the most alike were almost always near each other geographically, regardless of changes in environment. As Darwin studied the diversity of life in South America, he discovered species being replaced by very similar ones as he traveled along the continent, even into different climates.[3] Organic differences accrued with distance.

Darwin observed a similar pattern when he visited the Galápagos Archipelago and mused about this problem in his "Journal of Researches," now known today as *The Voyage of the Beagle*:

> It is probable that the islands of the Cape de Verd group resemble, in all their physical conditions, far more closely the Galápagos Islands, than the latter physically resemble the coast of America, yet the aboriginal inhabitants of the two groups are totally unlike; those of the Cape de Verd Islands bearing the impress of Africa, as the inhabitants of the Galápagos Archipelago are stamped with that of America.[4]

Why, Darwin asked himself in his journal, would this be? Why were the taxa of Galápagos "created on American types of organization?"[5]

Darwin, at the time, had already been exposed to certain concepts employed in evolution. His grandfather Erasmus, who had died before his birth, had believed that all species were related, and the French scientist Jean-Baptiste Lamarck had argued that species were mutable, changing from one form to another. But neither had provided persuasive evidence for the view. On his trip around the world, Darwin realized that the only way to answer the biogeographical questions with which he was confronted was not with natural theology but through a theory of evolution. Two species of the same genus that appear so much alike are related through a recent common ancestor, and this ties them to a common birthplace. The dominant distributional pattern of life on Earth, which suggests a general proportional relationship between taxonomic and geographical distance, is precisely what you would expect if all organisms were descended from the same ancestor. But the theological viewpoint, unconstrained by the unifying and co-locating processes of heredity, could not explain this distributional pattern. In *On the Origin of Species*, Darwin returned to the question of Galápagos and Cape Verde organisms and answered the problem he had stated many years earlier in his journal:

> The inhabitants of the Cape de Verde Islands are related to those of Africa, like those of the Galápagos to America. I believe this grand fact can receive no sort of explanation on the ordinary view of independent creation; whereas on the view here maintained, it is obvious that the Galápagos Islands would be likely to receive colonists, whether by occasional means of transport or by formerly continuous land, from America; and the Cape de Verde Islands from Africa; and that such colonists would be liable to modifications; the principle of inheritance still betraying their original birthplace.[6]

While this observation and explanation may seem prosaic today, a strong expectation remains, even among many educated people, that certain types of habitats—swamps, deserts, rainforests, oceanic islands, frozen tundra—should usually contain the same types of plants and animals, no matter where they occur on the globe. For example, a well-known commercial, shown in America every year at Christmas, uses the magic of computer-animation to depict a baby emperor penguin offering a drink to a polar bear cub. This fictional mingling of polar bears

4

with emperor penguins is actually quite common in the American media, often appearing in cartoons, Christmas specials, billboards, or magazine advertisements. In reality, polar bears live within or near the Arctic Circle while emperor penguins live on the opposite side of the Earth in Antarctica, and the two species have never naturally interacted. Both polar bears and emperor penguins do boast superb adaptations that equip them for a harsh and freezing existence,[7] but that does not mean they should appear everywhere icy. Polar bears are descended from the northern hemisphere brown bears, which include the North American grizzlies and Kodiak bears of the islands of Alaska. The domain of grizzlies extends so far north that grizzlies and polar bears have successfully mated in the wild, a union that produced an odd-looking bear, mostly white with brown patches, which fell victim to a hunter in the Canadian Arctic in 2006. Again we confirm Darwin's observation of one species replacing an extremely similar species from an adjacent region.

Likewise, emperor penguins descend from a southern hemisphere penguin-like ancestor, and other kinds of penguins are widely distributed throughout the southern end of the world, with some occurring as far north as Galápagos. The king penguin, the closest relative of the emperor penguin, lives nearby on a number of surrounding sub-Antarctic islands. Penguin-like fossils have been found, as one would expect, in the same southern regions as their current distribution: Antarctica, New Zealand, southern Australia, southern South America, and South Africa. Given their moderate but not exceptional dispersal abilities and the tropical climates separating the poles, neither identical species of polar bears nor identical species of emperor penguins could occupy both the Arctic and Antarctica. One perhaps could imagine in the distant future, some members of the penguin group, of which there are currently seventeen species, slowly expanding their range through the Americas so that they eventually reached the Arctic, but this process would still not produce identical or even most closely related species in both regions. In this hypothetical scenario, those penguins that slowly extended their range up the coasts of South and North America would have to adapt to the different climates and conditions of each new region, allowing them to colonize and spread further north. This would result in a number of new penguin species found along the coasts—like, say, the "tropical penguin," then the "Californian penguin," then the "Canadian penguin," finally an "Arctic penguin." But these hypothetical "Arctic penguins" would be the sister species of the hypothetical "Canadian penguins," and they would not be and could not be the most closely related species of the emperor penguins of Antarctica. All moderate and poor dispersing taxa that have adapted to such severe high latitude regions near the North or South Poles must have evolved from closely related organisms from the more temperate, lower latitudinal regions nearby. So Antarctica and the Arctic, given the current climatic barriers of the Earth, could not exclusively share the same species or even the two most closely related species of a genus unless its individuals were peculiarly proficient at long-distance travel.

Popular biogeographical mythology, like the fictional grouping of emperor penguins with polar bears, often springs from this same impulse to associate taxa with a certain kind of environment rather than a particular place. Even after I was familiar with the theory of evolution, I used to have similar distributional misconceptions regarding the Hawaiian Islands. Glimpsing pictures and videos of its rainforests and waterfalls, I had a hazy idea that the Hawaiian biota was vaguely Amazonian, peppered with an exotic and colorful assortment of lizards, frogs, and large branch-hugging snakes. Most of all, I suspected the islands were plagued with an intolerable number of mosquitoes and ants. I was very surprised to discover that, in reality, Hawaii has no native ants, mosquitoes, frogs, lizards, or snakes at all. Hawaii, in fact, completely lacks all native terrestrial vertebrates—no amphibians, no reptiles, and no mammals except for one species of bat. Those few vertebrates or ants or mosquitoes that do appear on Hawaii today are the result of recent human introduction. This should not really have been surprising as the Hawaiian Islands formed in the middle of the Pacific, 2400 miles (3900 km) from the nearest continent, which is far too distant to permit natural immigration of all but the most adept over-water travelers. Yet I still remember my astonishment that such a rich ecological system, so spectacularly diverse in its birds and flora, could be so poor in things that walk and crawl.

Darwin, as his writings make clear, suffered similar biogeographical expectations and was greatly amazed to find these expectations wanting. In his journal, Darwin noted that Galápagos strangely lacked frogs—as, in fact, did most of the islands he had been to—even though island environments seemed peculiarly suited for the amphibian life. Darwin knew that on some islands where frogs had been introduced, they almost always flourished. So why wouldn't frogs naturally occur on islands? Once again, Darwin returned to a question he had asked himself in his journal and answered it in *On the Origin of Species*: all frogs descend from a common mainland freshwater ancestor, and nearly all species die very quickly when immersed in salt water due to their water-permeable skin. Frogs do not appear on remote oceanic islands like Galápagos or Hawaii because they are particularly confined by large marine barriers, and the same is true, as Darwin wrote, for terrestrial mammals:

> Why, it may be asked, has the supposed creative force produced bats and no other mammals on remote islands? On my view this question can easily be answered; for no terrestrial mammal can be transported across a wide space of sea, but bats can fly across.[8]

The types of plants and animals that Darwin found on islands—and those he did not find—made it clear that taxa were not merely placed there by an invisible hand.[9] All island flora and fauna had to reach their oceanic residences under their own power, leaving the nearest continental region and their closest relatives behind. This is why all of the birds, insects, spiders, and plants that inhabit remote ocean islands exhibit certain traits that make them particularly amenable to long-distance overseas travel. Palm trees so frequently adorn islands because coconuts float and are extremely resistant to seawater. Wind carries the dust-like spores of ferns into the remotest interiors of the oceans. Long-jawed, orb-weaving spiders, which have the extraordinary distinction of occurring on every habitable landmass, can take to the air, gliding for great distances on their slender silk strands, drifting like plankton of the skies.[10] Given enough time, those organisms well suited for such transmarine jaunts eventually reach newly forming islands and successfully colonize them. Volcanic rocks that sprout from beneath the waves, at first lifeless and forbidding, soon begin to green and buzz and chirp. While these Magellans of the organic world continue to populate oceanic islands, those not disposed to transport by wings, wind, or waves must remain confined to the continents.

Consider, now, how compelling—how informative—these facts of distribution would be to someone trying to decide between the natural theology of pre-Victorian biologists and some other, less supernatural explanation for the origin of species. Island distributions led Darwin to realize that the extraordinary species of the Galápagos Islands—the bizarre and "hideous" marine iguanas, the giant tortoises, the peculiar finches—must be the result of a certain number of individuals reaching the islands from nearby regions like Central or South America or the Caribbean. He also knew that these Galápagan species all possessed unusual adaptations that differentiated them from the American type. Clearly, something had happened to the American species after they had reached the islands that had transformed them into the unique and curious forms observed on Galápagos today.

Moreover, each of these Galápagan groups comprised multiple species, like the thirteen species of finches or the two groups of iguanas—both land and marine. Why did none of these species also appear in South America? Was it conceivable that individuals from all thirteen species of finches managed to reach these isolated islands from South America due to different random events,[11] miraculously keeping all members of these similar groups in proximity? Was it reasonable to believe that after the thirteen similar groups of ancestors reached Galápagos from South America, all the other members of their species then died out on the continent? Certainly, the only reasonable way to explain this distribution would be if individuals from just one American finch species colonized the islands, and the thirteen resident species are all descendants of this first and only group of finch pioneers. Species, Darwin had started to realize even before he had finished his voyage, must have evolved.

Unfortunately, such powerful biogeographical evidence for evolution rarely appears in textbooks or renewed debates over

Creationism, now repackaged and stamped with the label, "intelligent design." This is particularly troubling given how much more powerful and explanatory the theory of evolution becomes within a biogeographical framework. While, like most people, I had grasped the basic principles of evolution in middle-school biology, I never had a truly profound and complete view of how evolutionary processes operate in the real world until I also understood it biogeographically—until, that is, I had a glimpse of the rich, global picture that impelled Darwin toward his discovery. In my view, biology teachers first describing the theory of evolution should place such evidence at the top of the syllabus, for once a student understands why there is no such thing as a Tahitian raccoon or why Koala fossils can only occur in Australia or why Arctic foxes cannot simultaneously inhabit both polar regions, he or she may begin to sense the full power of the theory of evolution.

The British geneticist J. B. S. Haldane once famously said that he would give up the theory of evolution if a fossil rabbit were found in the Precambrian. Haldane's quote shows, in an understated way, just how constraining the predictions of the evolutionary view really are. The theory requires the general emergence of organic complexity from simpler systems in a relatively predictable way—single-celled organisms, multicellular creatures, then aquatic vertebrates, all followed by the first lungfish-like creatures that crawled onto the land. This, in turn, had to precede the appearance of primitive four-legged terrestrial vertebrates and their eventual division into amphibians, reptiles, and mammals. Only after the first appearance of mammals could we expect to see the emergence of primitive fossil rabbit forms and finally modern rabbits. That the chronological order of fossils, implied by stratigraphic layering and reconfirmed by radiometric dating, should so precisely match the chronological order necessitated by evolution can be no coincidence.

Just as evolutionary processes require a continuous genetic flow through time, with no inexplicable disjunctions, those same processes also require a continuous genetic flow through space. For a botanical metaphor that may help enliven this perspective, we may imagine, instead of the tree of life drawn alone on a blank page, an intricate system of creeping ivy steadily extending and branching over the surface of the globe, tracing with its myriad stems the divergent lines of descent through continents and oceans. Two of the stems of this unifying global ivy, corresponding to a couple of primitive marsupial lines, would extend into Australia and divide into nearly 200 branches representing kangaroos, koalas, bandicoots, and all the pouched animals down under. In South America, we could trace a lone stem for Darwin's finches running from the coast into the Galápagos Archipelago and there splitting into thirteen branches that extend among the different islands. As all the stems and branches of this symbolic plant must remain whole and return to the same root, we can get an idea of what is and is not biogeographically feasible, helping to provide distributional analogues to that impossible Precambrian rabbit.

To cite a few examples, biogeographers could say, with similar Haldane-like boldness, that we would give up the theory of evolution if anyone discovered giraffes in Hawaii or fossil kangaroos in Spain or another group of Galápagan iguanas in Lithuania, each of which would represent an unacceptable spatial disjunction—a rend in the evolutionary creeping ivy that would be every bit as problematic as the temporal disjunction described by Haldane. As with the chronological order of fossils, the continuous genetic flow of evolution so constrains the possible locations

11

of plants and animals and allows so many opportunities for fal-
sification that we may disregard any theory that cannot explain
why organic distributions, all over the globe, should so persistently conform to such a faultless hereditary structure. The fact is
that no theory other than evolution can describe how the creeping ivy of organic descent could so envelop the planet yet always
remain whole.

A spectacular example of this organic bond moving through
space may help clarify. Those who hike or tour along western
North America, if looking for it, will notice certain species giving way to similar species in nearby regions—just as Darwin
observed in South America. The majestic redwoods—the Giant
Sequoias of Western Sierra Nevada and the Coast Redwoods
of California—provide one example. But the biologist David B.
Wake of UC-Berkeley has discovered a far more interesting case
involving a shiny black and bright red *Ensatina* salamander that
inhabits Southern California. As you steadily move north up the
Californian coastal mountains along the western edge of the San
Joaquin Valley, the populations of this salamander grow darker
and duller, even though the various adjacent populations continue to interbreed. In Northern California, at the northern edge
of the San Joaquin Valley, where the dullest of the salamanders
live, we find that the range of the salamander now turns southward into the interior of California, running along the eastern
edge of the San Joaquin Valley. The genetic differences continue
to mount as you now trace the populations southward along
the inland mountains, and the salamanders begin to change
color and display large dark splotches. Eventually, we return full
circle to the southern edge of the San Joaquin Valley and find
yellow-and-dark-splotchy salamanders from the eastern side of
the valley meeting the shiny-black-and-red colored salamanders

on the southern end of the valley. While every pair of neighboring populations interbreed as you encircle the valley, the genetic differences eventually become so great that once you come full circle, the two types of southern salamanders do not successfully interbreed. They have become reproductively isolated and meet the criteria for designation as separate species according to the biological species concept.[12] These salamanders are an example of what is known in biogeography as a *ring species*.

For additional insight, it may be helpful to describe this salamander ring in another way: by considering the two essentially distinct species in the southern side and tracing both lines along the opposite sides of the valley. As we follow these lines north, the two apparently distinct species, the eastern ones that are yellow with dark splotches and the western ones that are bright-black-and-red, continue to approach each other genetically, looking more and more alike, until finally they merge into the same species on the northern edge. Molecular analyses of the salamanders have provided a reasonably clear view of its evolutionary history. The shared ancestor of both the eastern and western populations began in the north. As the salamanders expanded their range along opposite sides of the valley, they began to diverge due to different selective pressures. Along the eastern side, the dark spots of the salamanders helped provide camouflage, while on the western side the bright reddish hues resemble the coloring of a local, poisonous newt that many predators know to avoid. Although the different subpopulations can interbreed, the gene flow among the salamanders is not persistent enough to prevent adaptation to the different habitats around the San Joaquin Valley.

The history of the salamanders may seem unique, but, excepting the peculiarity of the distributional ring, it is a relatively standard evolutionary tale. The reproductive isolation of the east valley salamanders from the west valley salamanders have resulted, through natural selection, in the diversification into two species. The only reason this example seems particularly compelling is because so many of the intermediary forms have done us the favor of surviving. The groups were also nice enough to bring the two most divergent end products of evolution back together again for a direct comparison.

Some people who deny evolution like to point to gaps in the fossil record, claiming no evidence suggests one species can evolve into another. When we find fossils that are either ancestral or directly intermediate between two forms—and, today, our museums brim with countless such fossils—the common retort is that gaps now exist between this intermediary and the existing species. Since we will never have fossils of every successive generation of a particular organism for tens of millions of years, one can always claim gaps. With the salamanders, however, we have no significant gaps. Each of the salamander subspecies can interbreed successfully with its closest neighbor—until we get to the two divergent groups at the southern end of the valley. The salamanders display small genetic changes, occurring through space, shaped by natural selection, resulting eventually in the development of two distinct species. While the evidence for evolution in the fossil record is overwhelming, it is biogeography that provides proof positive of the mutability of species.

As he traveled around the world, Darwin was able to observe this same geographical current of organic relationships, running from region to region. Today, such patterns are obvious to those who know to look for them: They confront us frequently and perpetually throughout the normal course of our lives, helping distinguish biogeography from nearly all other subjects. Most of us do not typically encounter fossils or observe the workings of embryology or come across many of the other well-known indications of evolution. Nor do we often meet with reminders of the efficacy of other scientific theories, say, plate tectonics or the molecular theory of gases or Maxwell's theory of the electromagnetic field. Most scientific models and principles follow from facts that require specialized knowledge and have been determined through careful investigation. But the distributional consequences of evolution are visible to practically everyone just through casual observation. Darwin noted in his summarial chapter in *On the Origin of Species* that some of the biogeographical facts he detailed "must have struck every traveler." But consider the effect such observations would have on those already familiar with his arguments. No evidence will ever fetch the true believers, but the probative value of distributions may at least help enlighten a few now bewildered by the intelligent design debate. To people familiar with biogeography, every visit to a national park or vacation in the Caribbean or hike in the mountains will continue to provide new and vibrant testimony to the Darwinian bond between life and Earth. The theory of evolution thus no longer remains distant and abstract; it becomes integrated into one's world view in the same visceral way that the mechanics of anatomy become a part of a surgeon.

As the facts of heredity demand, all organisms on this planet are physically linked through a material stream of genes that has flowed without interruption for billions of years from a single ancestral source. Those who truly grasp this enjoy a world decorated by striking floristic and faunal patterns, all serving to illuminate the secret behind life and the grandeur of Darwin's discovery. Those who deny evolution move through a fractured and irrational biotic world, devoid of organic relation and marked only by chaos and dim miracles. We should endeavor to close this gap. Biogeography, once a secret delicacy enjoyed only by geniuses, must now be elevated from its current obscurity and placed alongside literature and history as an indispensable component of a truly enlightened education.

By the time Charles Darwin had returned with HMS *Beagle* to Falmouth, England, he had begun to develop the view that all the world's flora and fauna had descended from a few or perhaps even just one common ancestor and that species evolved from similar pre-existing forms. As noted, this view was not original to Darwin, and others, including his grandfather Erasmus, had already suggested similar ideas. His problem was that he still did not know the mechanism. He knew that organic distributions suggested that species evolved; he just didn't know why. He would now spend much of the next two years of his life using all the available resources in England in an extraordinarily industrious attempt to find this last piece of the evolutionary puzzle.

His primary focus was on collecting facts, conducting analyses, and talking to experts about the great diversity in cultivated plants, animals, and birds, for he suspected a causal link between variations among domesticated breeds and variations in wild species. His first problem was confirming that many, if not all, of the great varieties of each cultivated group had descended from a common ancestor. Interestingly, this was not a well-accepted fact in Darwin's time, for some experts had argued that each breed of dog or sheep or pigeon had its own wild ancestor, eventually tamed by man. Darwin once again used biogeography to dispel that viewpoint.

In *On the Origin of Species*, Darwin pointed out that this view—that each particular breed of domesticated creature came from its own feral progenitor—would require the existence of dozens of different endemic species of wild cattle and sheep distributed throughout Europe. One author, Darwin noted, claimed the existence of eleven different wild species of sheep indigenous to Great Britain alone—one for every variety of sheep found on British farms. Yet, Darwin only knew of one mammal in Great Britain that was peculiarly British, and that France, Spain, and the other nations of Europe also had very few endemic mammals. Was it really feasible that in the recent past, each nation of Europe had been home to a multitude of unique ancestors to each of its various breeds of sheep, cattle, and dogs—and that they now had all disappeared? Moreover, some of the ancestral types were very difficult to imagine surviving in the wild, leading Darwin to ask, "Who can believe that animals closely resembling the Italian greyhound, the bloodhound, the bull-dog, or Blenheim spaniel, etc.—so unlike all wild Canidae—ever existed freely in a state of nature?"[13]

Darwin's argument needs no more additional support than that provided by the cartoonist Gary Larson in a now-classic *Far Side* comic. The single-panel cartoon depicts a number of small, meek, overly coiffed show-poodles, lounging beneath a tree, staring at giraffes in the distance. Cleaned bones and ribcages of prey surround them. Beneath the panel runs the title, "Poodles of the Serengeti." The absurdity of this juxtaposition helps underscore the artificiality, the total un-wildness, of many breeds of dog. The cartoon also lends some credence to H. L. Mencken's dictum that "One horse-laugh is worth ten thousand syllogisms."

Today, of course, we know that Darwin was right. Poodles, terriers, and boxers did not descend from three different natural ancestors. Instead, all the current breeds of dogs, from Chihuahuas to Great Danes, descend from wolves. And the same is true for most other domesticated plants and animals, each of which descends from a single wild ancestor.

More importantly, Darwin also knew what had brought about the variations within these cultivated flora and fauna. Many agriculturists had already written books on the subject and documented their progress in creating new varieties by the simple practice of allowing only those individuals with certain valued traits to reproduce. Darwin had collected many research passages that referred to the incredible plasticity of animals and plants, allowing farmers and gardeners to mold them, through years of selective breeding, to novel forms never before seen. In *On the Origin of Species*, Darwin refers to the veterinarian and breeding expert William Youatt, who "speaks of the principle of selection as 'that which enables the agriculturist, not only to modify the character of his flock, but to change it altogether. It is the magician's wand, by means of which he may summon into life whatever form and mould he pleases.' "[14] Darwin even noted that selective breeding often occurred unintentionally, as when a gardener simply yanked up the inferior plants or when a farmer ignored the least preferable of his animals when it came time to breed.

Realizing the importance of selection as a powerful transformational biological force, Darwin now merely had to determine how something like selective breeding could occur naturally. In October of 1838, at the age of 29, Darwin read Thomas Malthus' *An Essay on the Principle of Population*, and it gave him the missing piece of the puzzle. Malthus' work warned of the harsh struggle for existence that would accompany overpopulation, leading to famine, riots, and chaos—and placing a grim ceiling on population growth. Darwin recognized that this same struggle for existence in the wild would lead to a natural selection that would, like the gardener's hand, weed out less fit forms. Those fortunate individuals of each species that displayed certain traits that would give them a greater chance of survival and reproductive success than others would produce more offspring in the next generation than those not so fortunate. This process of natural selection would lead to slight changes in populations as certain advantageous traits would become more frequent from one generation to the next. These small changes would continue to accumulate in the vast stretches of time, leading to new species and, eventually, the wondrous biodiversity we observe today.

CHAPTER 2

The *Mesosaurus* Problem

How fossil locations helped revolutionize
our view of Earth history

Many people today view scientists as somewhat timid, pedestrian, socially clumsy homebodies. In "Socializing the Engineer," a chapter-section from *Genius: The Life and Science of Richard Feynman*, James Gleick detailed efforts at MIT to bring about the "socialization of this famously awkward creature," describing the young Feynman "and others like him" as "socially inept, athletically feeble, miserable in any but a science course ... so worried about the other sex that he trembled when he had to take the mail out past girls sitting on the stoop."[1] And, perhaps, the typecast is not wholly undeserved. Even those of us who do not work in a scientific field are certainly familiar with science professors and science majors. Many are often observed locked away in offices or at their dormitory tables, hunched over in the glow of a desk lamp, spending endless hours on the computer and politely refusing invitations to parties, pub crawls, and mud football games.

Yet the three men who are the focus of this chapter—Alfred Russel Wallace, Alfred Wegener, and Alexander du Toit—as well as the man just described, Charles Darwin—all share something in common that may serve as the counterpoint to this stereotype. Aside from being biogeographers, they were also grand and intrepid adventurers. Swashbucklers, all.

Alfred Wegener, the father of continental drift, at one time had *two* dashing and audacious world records to his credit. For the purposes of studying meteorology, Wegener had taken up hot air ballooning and in 1906 stayed aloft for more than 52 hours, longer than anyone else before him. Six years later, in 1912, he and his team struggled to survive an expedition in Greenland where they made what was at that time the longest crossing of an ice cap in history. Wegener also fought in World War I for Germany's Queen Elizabeth Grenadier Guards and was shot in the arm. After returning to duty, he was then shot in the neck—which ended his soldiering.

Another important combatant in the continental drift revolution, Alexander du Toit, also had a venturesome spirit. In the early part of the twentieth century, he helped map the Cape of Good Hope and traveled extensively in South America, India, and Australia.

Darwin's around-the-world journey has already been discussed in Chapter 1, but it is perhaps the overseas adventures of Alfred Russel Wallace, the co-discoverer of evolution, that were the most thrilling. He was one of the first Europeans to explore Rio Negro in South America, and he lived for a few years on the Amazon. When returning to England, in 1852, his ship caught fire and sank, taking Wallace's notes and his rare collections of plants, birds, and animal skins to the bottom of the Atlantic. Wallace and his shipmates ended up stranded in leaking lifeboats until their rescue ten days later. Still, Wallace would later travel to Indonesia, becoming one of the first Europeans to live in New Guinea for an extended period of time. Perhaps the most instructive way to think of these revolutionaries is not as scientists with boat tickets but as adventurers with journals.

The link between bold natures and biogeographical impulses even continues to this day. Jared Diamond, the biogeographer and author of *Guns, Germs, and Steel*, which will be discussed in Chapter 7, is also widely traveled and has spent much of his life living among hunter-gatherer societies of South America, southern Africa, Indonesia, Australia, and New Guinea. Gonzalo Giribet, a Harvard invertebrate-biology professor, is another plucky analyzer of organic distributions. He has not only traveled to all the continents to hunt specimens for biogeographical research, he also has Wegener's penchant for daring hobbies. He managed seventh place in the 2006 wind-surfing world championship in Spain and is an enthusiast of regatta-style sailing races.

The long-standing tradition of the adventurous biogeographer may offer another clue as to why the subject has produced so many revolutionaries. It is certainly not a coincidence that the writings of heretics like Wegener, Wallace, and Darwin are every bit as intrepid as their life histories. Wegener's *The Origin of Continents and Oceans* and Darwin's *On the Origin of Species* are not textbooks; they are grander, freer, more rugged. They do not smell of the classroom, they smell of rivers, beaches, swamps, and jungles. The risk-takers who wrote them are not the type to be daunted by professors or swayed by conventional wisdom. There is not a schoolmarm among them.

In other words, it takes a certain sense of audacity to attempt to revolutionize a science, to take on the full force of world opinion. And this may provide another reason why the study of plant and animal distributions has been so significant in the history of science. The subject of biogeography not only had all the facts required to overthrow centuries of orthodox thought in a number of fields; it also was the hobby of world travelers and so attracted those personalities who were not afraid to lead the fight.

If an extraterrestrial were to land in my backyard and ask to be taken to the most significant achievement of the human race, I would very likely lead the alien not to the Louvre or to Giza or to London or Manhattan, but to my bookshelves, and I would hand him *On the Origin of Species*.[2] So colossal in its significance, so sweeping in its explanatory power, so painstakingly erected upon countless empirical facts, *On the Origin of Species* stands as the Himalayan triumph of the human race. But, unfortunately, the bright glare of Darwin's masterpiece tends to obscure the importance of Alfred Russel Wallace, who had independently discovered the theory of evolution but whose name, like Elisha Gray, the co-developer of the telephone, is ever consigned to foot-notes and parentheses.

Interestingly, the path that led Wallace to the secret of life was eerily similar to that of his more well-known counterpart. Like Darwin, Wallace was a globetrotting naturalist, who explored South America. Like Darwin, he was an inveterate biogeographer who observed the global organic bond flowing through space and time and came to the same conclusion. Finally and perhaps most incredibly, like Darwin, Wallace happened upon the idea of natural selection as he reflected upon Malthus' *An Essay on the Principle of Population* and particularly its discussion of the struggle for existence.

Thus, while Darwin's trek toward scientific immortality may seem so personal and distinctive, Wallace duplicated it in all important details. This should not be too surprising because the theory of evolution was not an imaginative invention, but a fact of life waiting to be discovered. It seems reasonable that two observant men, particularly adept at pattern recognition and confronted with the same telling distributional evidence, would happen upon the same finding. Or perhaps we can put it another way. The discovery of evolution had to wait until some genius could combine an encyclopedic knowledge of biogeography with a recognition of the struggle for existence.

I have saved a discussion of Alfred Russel Wallace and his trip to Borneo for this chapter because it so naturally leads to the continental drift revolution. Wallace, more so than Darwin, directly tied evolution into geology. He was the one who first noted that life and Earth evolve together. It was after his trip to South America that he ventured to the Sarawak region of Borneo—and it was there, prior to his discovery of evolution, that he recognized what is now known as the Sarawak Law. In his 1855 pre-evolution paper, *On the Law Which Has Regulated the Introduction of New Species*, he introduced the tenet with the following observation:

> The facts proved by geology are briefly these: That during an immense, but unknown period, the surface of the earth has undergone successive changes....That all these operations have been more or less continuous, but unequal in their progress, and during the whole series the organic life of the earth has undergone a corresponding alteration.[3]

These observations led Wallace to put forth a biogeographical principle that needed explaining:

> **The Sarawak Law**: "Every species has come into existence coincident both in space and time with a pre-existing closely allied species."

This is truly the beginning of modern biogeography, the discovery that life and Earth evolve together. In the twentieth century, men like Leon Croizat, the brilliant and querulous bio-geographer who spent most of his years of study in South America, would expand on Wallace's view, noting the intricate connection between organic evolution, distribution, and geological events. And this in turn would escort many developments in modern biogeography. But it all really began with Wallace's little known paper.

Three years after his publication of the Sarawak Law, in 1858, Wallace would send his famous letter on evolution to Charles Darwin, suggesting that if Darwin considered it sufficiently interesting he should then send it on to the geologist Sir Charles Lyell. The letter Wallace wrote, "On the Tendency of Varieties to Depart Indefinitely from the Original Type," correctly identified the struggle for existence and survival of the fittest as the engine of evolution, a fact that would astonish Darwin and force him from his quiet, unpublished obscurity. Darwin, of course, had developed the theory by 1838 and had written to a number of colleagues about it, including Sir Charles Lyell, but he had never formally published on the subject. After Darwin forwarded Wallace's paper to Lyell, the geologist helped organize a joint reading of Wallace's letter and a few of Darwin's old communications on the subject before the Linnean Society—an arrangement that pleased Wallace. Despite the historic magnitude of this presentation, both papers were almost universally ignored. The theory of evolution would not become the subject of controversy until Darwin published *On the Origin of Species* in 1859.

Wallace also made another significant discovery while in Borneo when he came to notice a fascinating distributional divide now known as Wallace's Line—the most famous biotic barrier in the world.

Wallace's Line, as shown in Figure 1, connects two straits that run through the Malay Archipelago and, in a sense, identifies a divide between the Asian and Australasian biotic provinces. Or, as has often been said, this is the line that separates the realm of the tigers and pheasants from the realm of the kangaroos and cockatoos. While that helps give an idea, it would be more accurate to say that Wallace's Line is the very sharp start of an obvious biotic transition—the place where the ranges of an inordinate number of East Asian plants and animals, and particularly many placental mammals, suddenly ends. Similarly, the other line on the map, named after the biogeographer Max Weber, delineates the western boundary for many Australian and New Guinea biota—particularly the pouched marsupials. In between these lines is the biotic region Wallacea, which contains an impoverished mixture of Asian and Australasian flora and fauna that are better than average at crossing narrow marine barriers. As you move through Wallacea toward Australia, the flora and fauna become less Asian. As you move in the other direction, the plants and animals become less Australian. On Sulawesi (also known as Celebes), all that we find of the Australian marsupials is the little, tree-climbing cuscus, which lives alongside the placental descendants of those few mammals that have managed to reach Sulawesi from East Asia. These include a tusked fruit-eating pig, a dwarf buffalo, and four species of monkeys. No Australian marsupial has managed to get west of Wallace's Line, and except for a species of rat that invaded New Guinea and Australia five million years ago (and a number of bats, of course), no placental mammal has conquered Weber's Line.

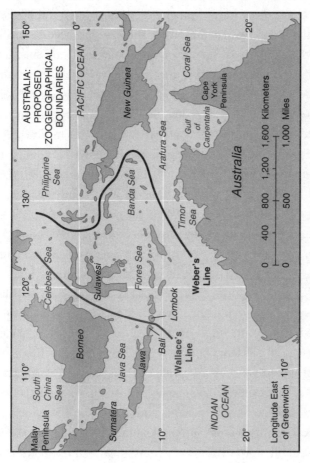

1. The positions of Wallace's Line and Weber's Line between Southeast Asia and New Guinea/Australia.

Wallace correctly deduced that the biotic line he found must correspond to a geological divide. The straits of Wallace's Line were deeper, he realized, and so far more persistent than the other marine barriers to the northwest. All the straits and seas that separate Java (also Jawa), Sumatra (also Sumatera), and Borneo from the Asian mainland are very shallow, and become fertile land when, during ice ages, the frozen poles impound such significant quantities of water it drastically lowers sea level around the world. But even during ice ages, the straits east of Bali and Borneo endured. In other words, as Wallace correctly surmised, his line marks the former southeastern coastline of East Asia. Weber's Line also has a similar geological rationale, running very near the boundary of the Australian continental plate. Thus, it is best to think of all of Wallacea—not just Wallace's Line—as the large and diffuse divide between Australian and Asiatic plants and animals—a series of saltwater sieves acting upon the terrestrial organisms of both regions that eventually filters out all the less adept travelers from moving between the regions.

Although I am aware of the drastic effect that marine barriers have on the organic ingredients of continents, I am still somewhat amazed by the narrowness of the often-impassable sliver of blue that separates Bali from Lombok—a mere 25 miles (40 km). This strait that serves as the western edge of Wallace's Line is often not distinguishable on a typical map of the world, yet it has proved impenetrable to so many plants and animals that its effect on the local organic composition was immediately discernible to Wallace, and, now, to all other biogeographers who have taken the brief boat trip from one side to the other. If such a slim thread of saltwater could have such an effect on the distribution of terrestrial taxa, imagine the biogeographical consequences of an ocean. This is particularly relevant to the continental drift revolution of Wegener because he discovered that, in the Mesozoic, the Atlantic and Indian Oceans seemed to have no effect on distributions at all. Fossil evidence indicated that numerous plants and animals passed between the regions on opposite sides of these oceans as if they were simply not there. Eventually, this discovery would help upend the view, believed by the majority of geologists and geophysicists even into the early 1960s, that continents are eternally motionless.

As with *On the Origin of Species*, Wegener's *The Origin of Continents and Oceans* contains a chapter devoted to biogeographical distributions. The chapter discusses many of the transoceanic disjunctions of many plants and animals, whether currently alive or long extinct, that biogeographers had been noting for decades and that now seemed quite perplexing given the evolutionary requirement of a physical biotic connection among all life. Wegener's list of these seemingly ruptured distributions included earthworms, parasites, reptiles, mammals, and various kinds of plants—all seeming to link Western Europe with Northeastern North America or Africa with South America or India with Madagascar. As Wegener correctly surmised, these regions had once been all part of the same continental landmass and had now drifted apart. Today the most famous disjunct taxa of the continental drift revolution, those routinely recited in elementary textbooks on plate tectonics are the seed fern, *Glossopteris*, and the early reptiles, *Cynognathus*, *Lystrosaurus*, and *Mesosaurus*.

The great southern continental landmass depicted in Figure 2 is referred to as Gondwana, named after a region in India where fossils of *Glossopteris* have been found. The Gonds were a people of Northern India, and Gondwana literally means

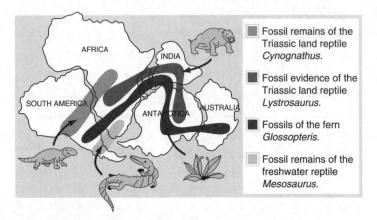

Fossil remains of the Triassic land reptile *Cynognathus*.

Fossil evidence of the Triassic land reptile *Lystrosaurus*.

Fossils of the fern *Glossopteris*.

Fossil remains of the freshwater reptile *Mesosaurus*.

2. The Mesozoic distribution of certain fossil taxa on Gondwana. The shaded bands represent fossil locations, but it is important not to be misled. The fossil discoveries do not really fall along narrow strips that run directly off the end of one continent and suddenly appear on the other. Instead the fossil sites are scattered about certain sections of the continents. These distributions provided extremely compelling evidence for continental drift because a number of the taxa, like *Mesosaurus*, were found nowhere else.

"land of the Gonds." Thus, the redundancy of the original term, *Gondwanaland*, has caused it to fall from favor. The fact that a *Glossopteris* site gave its name to the southern continental landmass helps emphasize that Gondwana was a biogeographical concept long before it was a geological one. And, as we shall see, although geologists, geophysicists, palaeontologists, and even some biogeographers remained unmoved throughout the first half of the twentieth century, the strength of this evidence is difficult to deny.

Glossopteris was a tree fern or perhaps a large woody shrub with leaves as much as 3 feet (1 m) in length. Its seeds do not float—and the possibility of birds carrying seeds across oceans, whether in their feathers or in their stomachs, can be ruled out because birds did not exist at that time. (Interestingly, trees have been around for 430 million years, but the first birds did not evolve until perhaps 150 million years ago. Thus, for the majority of their existence trees have swayed in the wind in relative solitude and silence, lacking nests, birdsong, or flutters of color.)

Mesosaurus was a freshwater reptile—one of the first to readapt to the life aquatic since reptiles first evolved from amphibians. It was tiny, less than 3 feet (1 m) in length, and had a long crocodile-like skull with nostrils on top allowing them to breathe without having to lift their heads from the water. The other two creatures, *Cynognathus* and *Lystrosaurus*, were both mammal-like reptiles, the group that would eventually spawn the ancestral species to all mammals. These thick, squat land creatures present similar distributional difficulties. We will discuss Gondwana and its breakup in greater detail in Chapter 4.

As is clear from Figure 2, the outlines of certain pairs of continents not only fit together, like South America and Africa, but the organic hues and tints of each of the landscapes match as well. When the regions are placed together, they reunite places decorated by similar flora and fauna—just as puzzle pieces fit together by both the shape of the outline and the various shades and colors of the overall picture. Wegener also had another piece of evidence, the textures of the regions as demonstrated by similar geological structures. This, Wegener argued, was mutually reinforcing evidence—evidence that dovetails with such precision that it is impossible to ignore.

Alas, Wegener, the meteorologist, was not formally trained in the subjects he was trying to revolutionize—and became the focus of hostility for many authorities in both geology and paleontology. The following comment by Dr. Rollin T. Chamberlin of the University of Chicago was fairly representative of the attitude that many established scientists felt toward this presumptuous outsider:

> Wegener's hypothesis in general is of the footloose type, in that it takes considerable liberty with our globe, and is less bound by restrictions or tied down by awkward, ugly facts than most of its rival theories.[4]

The irony of the statement is that the exact opposite was the case. Chamberlin's beliefs were the ones that were tenuous and speculative while it was Wegener and du Toit who were working with the extraordinary force of biogeographical patterns and so relying on unassailable facts. The conventional view of the time that continents were eternally motionless, no matter how firmly believed by experts of that time, rested on a great number of assumptions about which they could not plausibly claim to be certain. They had no compelling reason to accept as fact that they really knew everything that could happen deep beneath the crust and precisely what couldn't; that there was no possible internal energy source that could move continents. Their views had not been decided by evidence; they were merely familiar and long-standing.

Conservativism is one of the most important components in the process of scientific arbitration. Fantastic new theories or claims of incredible discoveries must be rigorously tested and empirically confirmed before finding acceptance. But it is also conservative to accept the necessary ramifications of incontrovertible data, even when it challenges traditional notions. What Chamberlin failed to realize was that his supposition that continents were eternally fixed was not a fact—but a scientific hypothesis that relied on a great number of suppositions about the unknown. It would have been far more conservative for scientists to accept that they did not yet know whether continents could move, for it was radical conclusion-jumping to assume otherwise.

But while geologists work in a field fraught with obscure and mysterious variables, biogeographers do not. The consequences of oceanic gaps on plants and taxa are simple and indisputable— despite the fact geologists and paleontologists of the time did dispute them. A little more than a decade after Wegener had died, this very subject became the focus of the David-vs.-Goliath debate between the drifter Alexander du Toit and the renowned George Gaylord Simpson. Simpson was one of the most influential and respected paleontologists of the twentieth century and a founder of the modern synthesis, which integrated the mathematics of genetics into Darwinian natural selection. He also incorporated the field of paleontology into evolutionary theory and was an expert biogeographer, showing the migrations of mammals between continents in the past. Still, in the early 1940s, Simpson wrote two articles attacking the biogeographical arguments for continental drift, particularly referencing du Toit's 1937 book on the subject.[5] Du Toit responded with an article, defending the original arguments. The back and forth comprises one of the most fascinating and important scientific altercations of the twentieth century.

Simpson attacked the biogeographical argument on two basic points. Some of the disjunctions, he argued, were the result of transocean voyages by rafting on driftwood or uprooted trees; others, like many of those between the southern continents, were the result of range expansion from Eurasia and North America. Eventually, Simpson believed, fossils indicating the northerly presence of these plants and animals would eventually be discovered. In other words, Simpson was arguing that it was twentieth-century geology, not biogeography, that was the settled science, that distributional patterns were not really that informative, and that spectacular dispersal events combined with fossil absences could lead to numerous misleading arrangements. Quoting Simpson:

> Some of the adherents to [mobile continents] complain this process [ocean-wide rafting, etc.], however possible, is improbable, that repeated improbabilities may amount to a practical impossibility, and that their opponents use this sort of distribution as a *deus ex machina* to solve all their difficulties. To some degree this is true; but the process is often less improbable than it seems.[6]

Simpson then argued that, given the vast stretches of time, we could still expect the improbable to happen. In other words, even if, for any given year, the odds of a clutch of eggs in driftwood or an Adam-and-Eve rafting pair successfully floating across the full extent of an ocean were a million to one, then over the course of five million years, we should not be surprised to see five full-ocean jaunts. Since this argument seems reasonable and since this is precisely what many scientists of the early twentieth century were willing to believe, it is important to understand why it fails to account for the fractured distributions discussed by Wegener and du Toit.

As we shall see in the next chapter, 50,000 oceanic islands provide 50,000 empirical tests—with each test lasting millions of years—that demonstrate precisely which types of plants and animals can cross narrow marine barriers, which can cross major oceanic gaps, and which are corralled by even the narrowest of straits. Recall the conspicuous distributional consequences of the ever-so-narrow Wallace's Line. Again, this is a barrier that has existed for millions of years, and countless non-aquatic plants and animals have been baffled by it. Yes, Simpson was exactly right about his math, but as we shall see, those miracle journeys that may occur once-in-a-million-years are actually how the *greatest* of dispersers conquer great marine gaps. Every few million years or so, bats, seals, ferns, and certain high-floating, lighter-than-air spiders will manage to travel significant distances across the ocean and reach their secluded and exotic residences. But the chances of the heavier land-dwellers successfully making such journeys, is, for all intents and purposes, nil, even over hundreds of millions of years.

But there is a biogeographical problem of deeper significance. Simpson's viewpoint not only necessitates the assumption that many of these poor dispersing taxa could actually make full-ocean jaunts, from say Africa to South America, it also entails that they then never went anywhere else. This is particularly problematic for disjunctions involving plants and animals that still exist today. As we shall see in Chapter 4, India and Madagascar share a number of groups of closely related freshwater fish and amphibians—both of which are particularly vulnerable to saltwater. If you are willing to assume this distribution is the result of their voyaging across the full extent of the Indian Ocean, then you are left with the mystery of why they have never managed to reach any other regions significantly closer—like any of the other islands in the Indian Ocean or, for that matter, Africa, which is merely separated from Madagascar by the Mozambique Channel.

Finally, recall that the biogeographical patterns caused by marine remoteness—the lack of walkers and crawlers on oceanic islands—became one of the facts that helped push Darwin toward the theory of evolution. Du Toit understood that Simpson's arguments left open the door for followers of creationism:

> The notion of random, and sometimes two-way, "rafting" across the wide oceans…evinces, however, a weakening of the scientific outlook, if not a confession of doubt from the viewpoint of organic evolution.[7]

Simpson's reliance on convenient fossil absences also had similar difficulties. If you try to use undiscovered fossils and the current continental arrangement to explain the exclusively South American and African distributions of Mesosaurus, you would have to assume the freshwater reptile actually had lived all throughout the rest of Africa, the Middle East, Asia, the Bering Bridge, North America, and South America,[8] but that their fossils have only been discovered in South America and South Africa. This is possible, but it is rather easy to show that such hypotheses become wildly improbable rather quickly. Let us say that there are five other fossil-bearing sites on the terrestrial route between South America and South Africa from the time period of interest—and the odds of a particular vertebrate group inhabiting any of these regions without leaving discovered fossils are 50% or 1 out of $2(^1/_2)$. (In reality, the odds of course would vary for each region and with each taxon, depending on countless factors, including the amount of exploration in the regions and the environmental circumstances that led to the preservation of the fossils. But this very rough estimate should help give an idea.) Then the odds that seven regions in a row would only leave discovered fossils in the two end regions, South America and South Africa, and supply no fossils in the five interim regions are $^1/_2 \times ^1/_2 \times ^1/_2 \times ^1/_2 \times ^1/_2 \times ^1/_2 \times ^1/_2$ or 1 out of 128 (1/128). You would then have to make this same improbable assumption for every other disjunct fossil organism—and multiply all those resulting fractions of improbability together.[9] Du Toit's response was correct:

> To argue that such southern disjunctive distribution is due to colonisation from the north through forms not yet discovered ... is neither scientific nor fair.[10]

The theory of fixed continents necessitates a conspiracy of these unlikely events—miraculous cross-ocean voyages and repeated examples of just right fossil absences—affecting only those plants and animals from the same pairs of regions that just happen to be connected on Wegener's now famous palaeomap.

The weakness of Simpson's biogeographical arguments may make it seem incredible that so many mainstream scientists were fetched by it. But Simpson's arguments certainly seem plausible when not carefully scrutinized, and, perhaps more importantly, the tone of his essays had the seductive haughtiness of an expert intimately familiar with his subject. He had authority and spoke with it, and when he began his essays against du Toit and continental drift with dismissive comments like the following, it must have been devastating:

> The fact that almost all paleontologists say that palaeontological data oppose the various theories of continental drift should, perhaps, obviate further discussion of this point and would do so were it not that the adherents of these theories all agree that palaeontological data do support them. It must be almost unique in scientific history for a group of students admittedly without special competence in a given field thus to reject the all but unanimous verdict of those who do have such competence.[11]

Such disdainful comments from one of the foremost authorities of the day certainly must have been intimidating to young researchers who might have started to open their mind to the possibility of Wegener's scheme. What finally ended all quibbles was the 1960s discovery of seafloor spreading and the extremely young age of all seafloor. As we shall see in Chapter 4, while the stable, continental cores are billions of years old, the oldest ocean floor is merely 200 million years old. Thus, the formation of the Atlantic and Indian Oceans in the past 200 million years was what broke the continents apart and pushed them around. This was the previously unknown phenomenon that had the force to move continents.

In 1912, Alfred Wegener wrote his first book detailing all the evidence for continental drift. In 1930, he died in a blizzard on an expedition in Greenland, more than 30 years before his theory would become conventional. Unfortunately, this is not that unusual a situation. Many scientists of the past who are acclaimed as revolutionary today—including John Herapath, J. J. Waterston, Ludwig Boltzmann—were ignored or attacked by conventional scientists of their time and, like Wegener, died unknown. Established academics have much invested in the theories they have taught to others and wrote about in books and journals, and this can, at times, lead to unreflective antagonism toward daring and mutinous ideas. The strength of the evidence advanced by Wegener and du Toit—the apparent obviousness of their claims, at least when looked at from our perspective—helps emphasize this point. The map showing the agreement of interlocking continental outlines with the repair of the biotic disjunctions would appear to end all debate. How could that have been a coincidence? Both men had gathered enough facts and evidence, presented the case in a sober and convincing fashion, and were attacked for it. The result was intellectual stagnation in geology, geophysics, oceanography, basic planetary science, and even biogeography for half a century—lost time on our scientific journey that can never be recaptured. But it was not the methods of science that misled us, simply the frailties of humans. In the end, the scientific process triumphed.

Those interested in intellectual revolutions would do well to focus on the plight of Wegener and du Toit, which serves as an archetype for the outsider's struggle against convention and offers another example of the effectiveness and even indomitability of the scientific method. Their tale also helps underscore the significance of biogeography to the history of science, exposing those features that make it such an unerring guide during times of darkness and contention.

CHAPTER 3

Pygmy Mammoths and Mysterious Islands

Why the geography of certain island groups,
like Hawaii, Galápagos, and the Channel Islands, has
turned them into a cauldron of evolutionary catalysts

"The black Lava rocks on the beach are frequented by large
(2–3 ft) most disgusting, clumsy Lizards. They are as black as
the porous rocks over which they crawl & seek their prey
from the Sea. Somebody calls them 'imps of darkness'. They
assuredly well become the land they inhabit."
—Charles Darwin's journal entry on his first sight of the
marine iguanas of Galápagos, September 17, 1835[1]

A lthough the thirteen finches and two groups of iguanas of
Galápagos were among the first creatures to submit to evolutionary explanations and serve today as symbols of Darwin's
great discovery, each group has harbored its own evolutionary
mystery that has not been fully explicated until recently. Before
discussing their closely guarded secrets and the evolutionary
forces that govern islands in general, it is important to clarify
the two mechanisms at work in evolution—heredity and natural
selection—which, in turn, unite the two elements of biogeography: biology and the environment.

As discussed previously, the various geological and climatic
effects that have differentiated the regions of the world and
greatly transformed the Earthly landscape are the driving forces

of natural selection. The same events that shape the Earth also mold plants and animals. Unfortunately, two misconceptions about this evolutionary dynamic have become extremely common. First, despite the popular view of evolution expressed in many venues, evolution rarely proceeds by giant mutational leaps—indeed, almost never—and second, evolution is not a completely random process.

Nearly all descriptions of evolution in popular film make the first of these mistakes—so frequently in fact that the notion of evolution by freakish mutation has attained the status of popular mythology: The movie and comic strip *X-Men* is about a race of human mutants, some good, some bad, who have all undergone superhero-like transformations. Some are wolf-like or lion-like or frog-like. One can control the weather; another can shoot lasers from his eyes. The mutants are supposed to represent the next step in evolution of mankind. In the film *Waterworld*, Kevin Costner plays a fish-like mutant with gills behind his ears—a marvelously convenient birth defect for someone on a planet now completely covered with water. The poster for the comedy-sci-fi movie *Evolution* featured a bright, yellow smiley face with three eyes.

While none of these movies would be confused with a serious documentary, all of these examples spring from and exacerbate the detrimental misconception that evolution is a theory based on extreme mutations. In reality, the changes that lead to speciation are typically so slight that they are essentially unnoticeable—even over the course of a few generations.

Consider the evolution of the marine iguanas of Galápagos—one of the two groups of closely related iguanas found on the islands. These most seaworthy of all reptiles have long, flattened tails and partial webbing at the base of their toes, both of which help provide propulsion as they serpentine through the sea. Their toes are long and strong, allowing them to grip boulders tightly amid crashing waves. They have blunt noses which aid in scraping the seaweed off the rocks, and they are able to collect all the excess salt they ingest in their nasal glands and then excrete it through white, explosive sneezes.

Yet all these adaptations are merely exaggerations of traits that appear in other iguanas. None of these qualities suddenly appeared fully formed in a single freakish hatchling, which then succeeded in taking over the species. Instead, the evolutionary story is far less fantastic: Not long after the South American terrestrial iguanas had reached those desolate East Pacific islands, they at some point began supplementing their diets by eating seaweed exposed on rocks at low tide. In more desperate times, perhaps when rising sea levels would shrink their habitats, the iguanas would venture a little below the surface to eat seaweed on the underwater rocks farther from shore. The natural variations observed among all species ensured that some of the iguanas had slightly more webbing between their toes or slightly flatter tails than others, which would confer a slight advantage to these seaweed-eating reptiles. Over generations, the genes for these marine traits would begin to increase in frequency throughout the population of iguanas. This process, endlessly repeated for millions of years, would lead eventually to the iguanas we observe today. Marine iguanas have now become so adapted to their seaweed-dining lifestyles that they can dive more than 40 feet (12 m) below sea level, and some sources suggest they can stay underwater for more than 30 minutes.

Perhaps, topiary, the art of sculpting giant shrubs into various shapes, provides the most elegant and useful analogy for the way natural selection shapes species. The bush tends to grow in wild and random directions, but the gardener, by pruning the unwanted branches, sculpts a particular shape out of the random growths. Likewise, species over the course of millions of years develop many slight, random variations, and natural selection prunes all deleterious changes, leaving only those that help a species survive and reproduce within its environment. Proponents of intelligent design who believe that certain seemingly just-right adaptations are too improbable to happen by chance are focusing exclusively on the biological and not enough on the pressures of the environment, which is to say, they are focusing on the shrub and ignoring the pruning shears. To understand biogeography, that is, the exact interplay between the biological and the environmental, one must keep in mind all the wild, random, branching mutations that had occurred in those island iguanas—and all of the other iguanid forms that would have been possible given different environments. Though blind and purposeless, natural selection—and there is no better word for this—*sculpted* the marine iguanas out of that wild, branching, mess of possible iguanid forms, clipping those other deficient stems and leaving only those traits that were beneficial for their wet and salty way of life.

Numerous iguanid variations are still hatching on Galápagos today. Anyone who observed the marine iguanas over generations would notice certain other deviations and offshoots—iguanas with shorter toes, iguanas with little to no webbing, iguanas with rounder tails or longer noses, all struggling to survive. In a different place and time, such variations may be very beneficial. But they are not currently helpful to the seaweed-eaters on Galápagos and so inevitably fall to Nature's scythe. Like topiary sculptures, such well-defined species that are so peculiarly suited to their environments have to be maintained, and natural selection is a constant and indefatigable pruner. This is why species seem so perfectly matched to their environment, and why some people, who are innocent of such explanations or feel ill-disposed to them, think the animals had to have been designed for a particular niche by a wise and invisible entity.

The relatively simple evolutionary story of the Galápagan iguanas has recently been confronted by a mysterious difficulty. A 1997 molecular analysis uncovered the fact that both the land and marine iguanas of Galápagos are actually much older than the islands themselves. The islands are roughly three to four million years old, and both genera of iguanas are roughly ten to twenty million years old. If you know a lot about evolution but only a little about the geological history of Galápagos, this fact should astonish you. At least superficially, the theory of evolution would seem to require that the two groups of iguanas, like any other species peculiar to an island group, would have to be younger than the islands they inhabit. After all, the island isolation of the iguanas, first from their South American relatives, then from each other, was what instigated the adaptive changes that made them special, which is to say, that made them species. Naturally, this would require that the iguanas' adaptation to island life would have to postdate the islands.

Is it possible the molecular clock analyses used to determine the iguanas' age was wrong? Well, it is possible, for while the science behind such investigations is simple, the conclusions rest on a number of assumptions. Molecular clock theory follows from the evolutionary fact that when two populations of a species are isolated, the DNA of the separated groups become increasingly differentiated as mutations accrue and the two populations begin to diverge. Many of these accumulated variations are of no help to chronological studies because they are influenced by natural selection, and so are subject to the vagaries of environmental pressures. A clock based on such differences would be neither steady nor predictable.

But other sections of DNA have no empirical effect on the fitness of the individual and so are not influenced by selective forces. In fact, certain sections of DNA, it seems, do not code for protein at all and so have no conspicuous biological purpose. This "neutral" DNA accumulates mutations randomly and does so without having any effect on the organism. So, while some controversy still surrounds the precision of molecular clocks, it is generally believed that changes in neutral DNA occur relatively steadily over time. It is possible the iguanid clock is wrong, but the estimates were careful and conservative, and 10 million years ago appears to be the latest that the two iguanid genera began to diversify.

So what is the explanation for the mystery? If the iguanas are at least that old then this implies the existence of other older islands that must now be submerged somewhere in the vicinity of Galápagos or between the islands and South America. And so indeed it has turned out. Galápagos sits over a "hot spot," a molten rupture on the seafloor that spews molten basaltic flows into the ocean, some eventually building seamounts so large they emerge, steaming, from the waves and become volcanic islands.

For many decades, it was believed these hot spots sit above long, narrow plumes of magmatic material that extend down deep into the mantle, far below the crust, and that these plumes remain stationary with respect to the Earth's axis. The oceanic plate then moves over the "fixed" plume, conveying the islands away from the hot spot. Recently, some researchers have challenged the simplicity of this model, with some showing evidence that some hot spots are not stationary, and a few others even arguing they are a much shallower phenomenon, acting more like a progressive tear in the Earth's crust rather than the product of a narrow and deep plume. But the seminal cause of hot spot islands is not relevant to our discussion, as everyone agrees about the timing and consequences of the process.

Hawaii is also a hot spot archipelago and has a similar history. Mauna Kea, the highest peak in the Hawaiian chain, rises some 5.5 to 6 miles (9–10 km) from the ocean floor, making it, from top to bottom, the tallest mountain in the world (although its upper regions are considerably more inviting than those of Everest). Progressively, every few hundred thousand years, newer mountain-islands have continued to form over the Hawaiian hot spot as the older islands to the northwest have continued to sink. The Lo'ihi Seamount, sometimes known as "the next Hawaiian island," is an active underwater volcano about 19 miles (30 km) southeast of the island of Hawaii. It is still about 0.6 miles (1 km) below sea level, but if it continues to build itself up with more eruptions, it will likely emerge from the waves within the next 100,000 years.

The pressures that this geological pattern of rising and sinking places upon the Hawaiian and Galápagan taxa are quite extreme, forcing them into a rather desperate struggle against time. All the flora and fauna on every island of these chains are eventually doomed to drown—or, at the very least, perish in an increasingly frenzied clash for the dwindling resources on a slowly shrinking habitat. The only species that will continue to survive on the archipelagos are those that are able to keep sending pioneer colonies to the islands recently formed. Those that fail to reach these young rocky isles of salvation will be lost to the sea.

Thus, Darwin's famous islands comprise the young, high, and exposed western tip of a much older and larger Galápagos archipelago, most of it currently underwater. This basaltic chain extends 620 miles (1000 km) eastward from the relatively young islands that Darwin visited to the much older underwater seamounts just off the western edge of the South American plate and the white beaches of Ecuador. The existence of this seafloor ridge has long been known, but it was not clear that any of the seamounts had ever been above sea level, as the molecular dating of the iguanas would seem to require. In 2003 scientists analyzed samples that they dredged from some of these underwater mountains and determined that islands along this Galápagos Archipelago have been continuously emergent for at least the last 17 million years, explaining why and where the iguanas had diversified before the formation of their current residence. As R. Werner and K. Hoernle, the two scientists who analyzed these samples, wrote: "These new data significantly extend the time period over which the unique endemic Galápagos fauna could have evolved, providing a complete solution to the long-standing enigma of the evolution of Galápagos land and marine iguanas."[2] The ten endemic species of flightless weevils of Galápagos also appear to have begun their radiation prior to the oldest current extant island and so have a history that appears to parallel that of the iguanas.

This is a particularly intriguing example of the intricate interplay between life and Earth because it shows how researchers can use biological methods to help illuminate the geological history of a region. More specifically, it shows that scientists, simply by studying the tiniest bits of an iguana, can actually discover evidence for the presence and location of previously emergent seamounts along the Pacific seafloor. Traces of their now-sunken island homes are implicated in the iguana DNA. It also confirms that the evolutionary history of the iguanas would have remained incomplete, even confusing, without fully understanding the intimate connection they have with the geological history of the Galápagos.

The other species on Galápagos that had become so important to the history of science, "Darwin's finches," have also recently surrendered a long-kept secret. The mystery of these finches involves "adaptive radiation," the evolutionary process by which a single ancestral animal or plant divides into a great number of species, all modified to fit different niches. The finches are a peculiarly ostentatious example of adaptive radiation. Finding little competition for resources, the original finch ancestor has now evolved into more than a dozen species—each conforming to some particular ecological nook or cranny. Some have smaller beaks now specialized for eating smaller, softer seeds; others have large beaks for larger, more difficult seeds. Some forage on the ground, and some remain in trees. Some even use tools, tiny twigs or cactus spines, to dig insects out of the bark. The small-beaked finches on Wolf Island have become quite aggressive eating eggs they have broken by rolling them off cliffs, or, like vampire bats, lapping the blood of seabirds from tiny wounds they inflict.

The more than fifty species of Hawaii's colorful small birds known as honeycreepers are another well-known case of adaptive radiation. The most famous of the group are those nectar-sippers that have developed long-curved bills to reach the sweet liquid at the bottom of elongated flowers shaped like slender vases. Scientists estimated that the birds, all descendants from a Eurasian rosefinch (or perhaps an American house finch), reached their island home within the past few million years and began speciating within the first few hundred thousand years of their arrival.

Adaptive radiation may seem like a different form of evolution, but it is really just the standard evolutionary tale. Isolated populations begin to evolve according to their different environments. But whenever a particular group reaches a new site with few predators and competitors, say, when finches colonize an island, they suddenly face a multiplicity of new opportunities. The environment is like a biological vacuum, and many of those plants and animals that seem particularly prone to speciation will tend to diversify and take advantage of innumerable potential niches. Mammals, in all their varieties, also provide an example of adaptive radiation—but a much broader, messier, and more complicated example. All mammals are descendants from the same ancestor, and they have radiated, dramatically so, to fit a great number of niches, particularly after the extinction of the dinosaurs provided a sparsely populated world to conquer—a global void with countless potential lifestyles waiting to be filled. Some mammals found ease and safety in the trees while others evolved for the life aquatic. Some became slow, large grass-eaters; others fast, quick-thinking hunters. Some fly, some burrow underground, some enjoy life on mountains. Dinosaurs have provided another elaborate example of adaptive radiation, branching into myriad forms after the Permian-Triassic extinction event 250 million years ago had emptied the earth of over 70% of all larger terrestrial vertebrates. Even the entire diversity of all life on Earth, starting from its murky, chemical beginnings to its current spectacular variety, is an example of adaptive radiation.

But the adaptive radiation that has occurred on island groups, and particularly Galápagos, is special. Island species that have generated so many new forms often flagrantly betray their Darwinian relationships and the secrets of their origins. In brief, adaptive radiation on islands is noisy, flamboyant, in-your-face evolution. It is evolution-out-loud. All the species of tortoises and finches of Galápagos, like those of the Hawaiian honeycreepers, have not only put their specializations on display, they have remained geographically confined with all their sister species, making those differences pronounced. Adaptive radiation is the reason why Darwin wrote the following about the Galápagan taxa: "Reviewing the facts here given, one is astonished by the creative force ... displayed on these small, barren and rocky islands."[3]

The questions are how and why? What was it about the Galápagos chain that had turned it into such a powerful factory of evolution, and what was it about the finches that made them so malleable to such selective forces? Why did they diversify over the course of a few million years into so many different species? Yes, the islands had all the amenities necessary, providing a platform for so many new, possible niches. Yes, these island chains comprise a group of independent laboratories, each with a slightly different climate, group of competitors, and food sources—all waiting to act upon any new group of inhabitants. But not all islands or all species are so conducive to adaptive radiation. What is the difference?

One of the most important factors in speciation is reproductive isolation, and island chains like Galápagos and Hawaii provide a significant number of opportunities for such isolation.

Galápagos has five major islands and numerous small ones, all available to colonize. Some are very close together, less than 6 miles (10 km) apart, others are more than 30 miles (50 km) away from the nearest major island. But varying sea levels, combined with their own patterns of erosion and subsidence, can severely change these distances. Recall that all these islands are the exposed peaks of an underwater mountain chain, so dropping sea levels can dramatically increase the size of the islands and decrease the gaps between them. The shoreline of the Big Island of Hawaii, which is now 30 miles (50 km) from Maui, was likely only 9 miles (15 km) from its island neighbor from 500,000 to 600,000 years ago, just as its oldest islands completed their shield development. Some plants, birds, insects, and reptiles particularly adept at negotiating the narrow gaps between the islands during times of low sea level will suddenly find themselves isolated on the different islands during high water stands. This can happen repeatedly over the course of a million years—as the marine barriers continue to expand and shrink.

Thus, the dispersal capability of an organism relative to the distances among the islands is an important factor in the likelihood of its isolation. For the most part, it is the moderate dispersers of island chains that are the most likely to radiate. Weak over-water travelers, of course, cannot even reach Galápagos or Hawaii. And the great over-water travelers have little difficulty moving from island to island, even when the distances are greatest, so genes continue to be shared among the populations, and speciation is prevented. The endemic tree fern, *Cyathea weatherbyana*—which, despite some superficial resemblances, is not a true tree at all, grows on the misty, volcanic summits of four of the larger Galápagos Islands. Ferns reproduce by spores that can float for many miles on the wind, so these various tree fern populations on the different summits are not really separated from each other. Their range does not consist of just a single island, but a larger Galápagan group.

Those plants, insects, birds, and reptiles able to cross narrow marine gaps frequently enough to reach all the other islands repeatedly—but not so frequently that they prevent speciation— are the prime candidates for adaptive radiation. Darwin's finches are just such a goldilocks-disperser, a bird whose propensity for over-water travel is "just right" for diversification. But they also have another important characteristic recently discovered by the husband-and-wife evolutionary biologists Peter and Rosemary Grant—a trait that may be the main reason why the birds have become such a storied feature of Darwin's discovery.

Since the early 1970s and continuing to this day, the Grants have spent six months of every year on Galápagos analyzing the finches—studying them in a way that perhaps no other creature has been studied. The scrutiny they have afforded the birds has been so detailed and precise that they have been able to associate directly certain environmental pressures with particular organic changes. For example, in 1977, a severe drought drastically reduced the edible seed supply on the small Galápagan Island of Daphne Major—at times leaving only the bigger, harder seeds and island weeds to feed the medium ground finch, *Geospiza fortis*. As a result, the population of the finch plummeted from 1200 to 180. The Grants' detailed measurements showed that the survivors were nearly 5% larger and their beaks roughly half a millimeter deeper and slightly longer than the average of the pre-drought populations. These were the individuals that were best able to survive a switch to a diet of larger and tougher seeds, the lucky few who could most easily crack them.

Among their many other discoveries, the Grants have also determined that the mating songs of the birds may have provided the most significant impetus in their remarkable radiation. According to their research, the finch chicks learn the mating songs from the father, and while the females never sing themselves, they choose mates, at least partly, due to similarity to the songs of their father. When one group of the species ends up on a different island, these songs, like their genes, begin to accumulate random variations. Very quickly, the songs among the two separated groups become so different that it inhibits mating among the populations when they are reunited. In other words, even relatively short separations of a finch population are all that are necessary to start them on the path to reproductive isolation and speciation. This dramatically increases the finches' opportunity for speciation.

Returning to the story of Darwin, it is interesting to note that even with all this, the thirteen varieties of finches that Darwin collected did not impress upon him while on Galápagos. Darwin did not even recognize many of them as finches. Instead, it was the tortoises and mockingbirds that had first pushed him to start thinking of their relationships. But within months after he had returned to London, the taxonomist John Gould inspected the bird collection and informed Darwin that these different species of small birds were all part of the same finch genus and that they were all exclusive to Galápagos. It was perhaps this fact—of all the information that he had gathered on his journey—that struck Darwin with the most force. In the second edition of his *Journal of Researches* (1845), fifteen years prior to his coming forward with *On the Origin of Species*, Darwin became brazen enough to start hinting at his discovery: "Seeing this gradation and diversity of structure in one small, intimately related group of birds, one might really fancy that from an original paucity of birds in this archipelago, one species had been taken and modified for different ends."[4]

Darwin's arrival on the rugged, black-lava landscape of the Galápagos Islands marked the final component in a series of events that brought one of our greatest scientists to one of the most telling places on Earth. The fiery and steaming birth of Galápagos, its age and isolation from South America, the distance apart of each of its islands, the dispersal abilities and propensities and even mating habits of a number of its strange creatures—all came together to expose the secrets of life.

But this still did not strike Darwin immediately. He did not, while exploring its rugged and alien landscape, see a tortoise or pick up an iguana and immediately grasp the origin of life's diversity. He was still, in essence, a Creationist, still very much a product of the pre-Victorian ideas and biases that surrounded him. But something on Galápagos had begun to poke and prod him. And at some point, within the months after his return to London, as he continued to reflect on his voyage and learn more about the specimens that he had collected, the significance of the island fauna suddenly became clear. It was on Galápagos that he had been closest to the secret. The archipelago, Darwin knew, was a relatively recent geological formation, thrust upward from the Pacific floor via endless eruptions of undersea volcanoes. And if the Galápagos Islands were relatively young, then all of these strange species peculiar to Galápagos had to be relatively young as well. Like a detective who had just visited a fresh crime scene, Darwin realized that something had occurred in the relatively recent past, and perhaps was still occurring, right there on Galápagos, that had brought these new species into being. By the autumn of 1838, he had his explanation, and once again, in his second edition of his *Journal of Researches* (1845) he provided another daring and allusive hint to that discovery: "Hence," Darwin wrote about Galápagos, "both in space and time, we seem to be brought somewhat near to that great fact—that mystery of mysteries—the first appearance of new beings on the Earth."[5] But by that time, Darwin already knew that Galápagos had not only brought him "somewhat near" the mystery, it conspicuously flaunted much of the evidence he had needed to solve it.

As noted in the Preface, oceanic islands typically have few *native* non-flying, terrestrial vertebrates—and those islands that are thousands of miles from the nearest continent, like, say, Hawaii or Pitcairn, have none. All creatures that live exclusively on land, have a backbone and cannot fly and are confined to the continents and nearby islands. Those walkers and crawlers that we do encounter on the remoter outposts, almost without exception, have been introduced by man.

I say, "almost," because I should, perhaps, except certain species of geckos and skinks from this general island rule as it is still unclear whether they have been introduced or have naturally rafted to some remote oceanic islands. For example, brown-tailed copper-striped skinks, a diminutive lizard that is thinner than your little finger and, discounting its breakaway tail, is not much longer, are found all over the Pacific from Fiji to Hawaii to Easter Island and essentially everywhere in between. Likewise, the Pacific islands are teeming with geckos, particularly the house gecko and mourning gecko. The gecko is another small lizard, often nocturnal with dull skin and bulging eyes, usually less shiny and sleek than skinks (the brightly colored "day gecko" being an obvious exception to the general gecko pattern). While many biogeographers believe that the presence of these lizards on remoter Pacific islands—and particularly east of Samoa—is exclusively the result of human introduction; others have supposed that some geckos and skinks reached a few of the remoter islands through rafting on driftwood or uprooted trees.

These tiny lizards certainly do have peculiar features that would seem to increase their opportunities for accidental and successful rafting voyages. The eggs of many geckos are saltwater resistant, and such eggs have been found in driftwood. Also, mourning geckos are parthenogenetic, which is to say, all of them are female and reproduce without sex. Thus, you don't even need the classic "Adam and Eve" pair to colonize the island. Instead, it is possible a single voyager could do it all by herself.

As is nearly always true in lizards, the parthenogenetic nature of mourning geckos is the result of hybridization, the mating of two members from two different, sexual species. Researchers have determined that such interspecies mating, likely on the Arno Atoll of the Marshall Islands, eventually produced four daughters who were able to reproduce through asexual cloning. These four founding reptile Eves then became the mothers of the four major groups of mourning gecko clones that have become sprinkled throughout the Pacific.

Parthenogenetics can have its benefits, the most obvious being that the creatures do not need to waste a lot of time and energy finding and courting mates or fighting off romantic rivals. The downside is that all of the individuals of the species are essentially identical, which is typically a disadvantage in the complex and mercurial world of behaviorally sophisticated vertebrates. Sex is a very active fuel for evolution because it shuffles up the genes from two individuals and deals them out again in novel ways. Sex leads to greater variety, and variety is what allows a species to adapt to changing and ominous conditions. In stressful times, those individuals who are superior at eluding predators or finding food will have more opportunities to disseminate their genes. In contrast, a group of clones, generally speaking, has no individuals that are either superior or inferior in confronting certain dangers or hardships. When one species learns to prey upon one clone of the group, it has learned to prey upon all of them. Successful adaptations among parthenogens must wait for chance and beneficial mutations. Yet if there were any place in the world that would be most hospitable to the all-female mourning geckos, it is the remoter Pacific islands, where they have had to confront few predators and had little to no competition.

While their small size and ability to clone themselves would seem to make the geckos prime candidates for rafting and colonization, these same qualities would also seem to make them eminently successful stowaways. And it is likely that most, if not all, geckos and skinks on remote Pacific islands are the result of their hiding on boats, particularly those of the first settlers of the Pacific islands, the ancient Polynesian voyagers. One particular tree-dwelling skink, *Emoia tongana*, has an extremely peculiar distribution, having been found on Tonga, Futuna, and Samoa. This is a distribution that is not matched by any other terrestrial vertebrate, but the biologists Christopher C. Austin and George R. Zug have recently pointed out the capital fact that their irregular domain coincides precisely with islands ruled by the Tongan monarchy of the twelfth and sixteenth to seventeenth centuries.[6] Tongan traders and warriors sailing between the islands on their large double-hulled canoes had probably ferried this inconspicuous passenger.

PYGMY MAMMOTHS AND MYSTERIOUS ISLANDS

Usually, we are able to infer today which plants and animals have been introduced to islands because they will not have had time to differentiate markedly—and so remain essentially identical to the species of the regions whence they came. Moderate dispersing taxa that managed to reach the islands through natural means millions of years ago will have had time to adapt to the island environment. So they become peculiar to that island or "endemic." But some researchers have noted that these little lizards often tend to be morphologically conservative, which is to say, they do not experience significant physical changes over time, even after speciation. Thus, what we may think is a well-known species of skink on a remote Pacific island may really be a "cryptic species"—a skink that rafted to the island a long time ago but whose subtle adaptations have not yet been noticed.

Regardless, we can at least state the following as a fairly hard rule of island biogeography: With the possible and perhaps even unlikely exception of the diminutive, ubiquitous, and biogeographically confusing geckos and skinks, the remoter oceanic islands are utterly devoid of indigenous, non-flying, terrestrial vertebrates. To put it in an even simpler way, on places like Hawaii, Tahiti, Pitcairn, Easter Island, etc., none of the natives have four legs—flippers possibly, wings certainly, but not four legs. The aboriginals that you encounter instead tend to be the gold-medal dispersers—certain types of land and sea birds, particularly terns and noddies; micro-bats are surprisingly good dispersers; certain kinds of spiders and insects; many types of ferns. Those effortlessly transported over the globe by wings, wind, or waves are the only natural remote-island residents.

The long-jawed, orb-weavers, known to biologists as Tetragnatha spiders, are rather special in this regard, appearing literally on every habitable landmass. One biographical analysis has shown that the Tetragnatha spiders from the remote islands of the Hawaiian, Marquesas, and Society archipelagoes are *not* most closely related to those from nearby island groups—as they would be if the spiders had used the islands as stepping stones. Instead, the resident spiders from each of these three far-flung archipelagoes arrived into the center of the Pacific directly from the continents. The orb-weavers of Hawaii, for example, seem to have invaded the island twice from America.

Here, now, is a taxon that appears to violate the biogeographical pattern that led Darwin to the theory of evolution. As opposed to the central Pacific orb-weavers being most closely related to the orb-weavers closest to them, they are instead the nearest relatives of spiders a great distance away. A careful analysis of the spiders' prodigious dispersal ability helps explain why.

Young Tetragnatha spiders are carried to new territories by a process called ballooning. Needing a new habitat (or more precisely—instinctively compelled to behave in a way that will bring them to a new habitat), the juveniles will crawl to the top of a plant or branch, lift their abdomen upward, and release a thin silken thread into the air. Wind then catches the bit of silk and carries the spider upward, often bringing them to another part of the field or forest. Sometimes, the spiders drift for tens of miles, or even hundreds of miles, and, at times, if they are carried high enough and catch a more powerful air-stream, even thousands of miles. Their fortune is entirely given over to Aeolus, the moody God of the Winds. A study of the aerial plankton collected by planes flying over the China Sea has determined that 96% of all arachnid floaters in this region are Tetragnatha spiders, confirming that the Pacific is indeed subject to a constant rain of these long-jawed arachnids. Hence, the extent and randomness of their distribution is certainly not surprising. Their allotment is so chaotic because their method of dispersal is so chaotic. So we must be careful to note that the overwhelming biogeographical pattern that shows a link between genetic and geographical distance does not necessarily apply to those taxa that continuously submit themselves to what amounts to a random-location generator: dispersal by lottery. If all organisms were subject to such persistent and chaotic scattering, it is likely that Darwin and Wallace would not have happened upon the theory of evolution as soon as they did—perhaps not at all.

As we get closer to the continents on both sides of the Pacific, we begin to run into the silver-medal dispersalists—creatures that are not able to reach the remote interiors of the oceans but, given the right placement of stepping stones, are able to venture fairly far into the seas. Lizards, tortoises, and rodents comprise this second tier of oversea travelers, followed by small carnivores. In general, the best reptile and land-mammal dispersers tend to be small and determined clingers and so are prone to natural rafting on uprooted trees. Tortoises are also efficient at crossing moderate gaps because they float and are extremely resistant to dehydration and starvation.

The poorest dispersers from the continents are amphibians and fresh water fish that have little to no saltwater tolerance, and the larger mammals. Imagine a giraffe trying to flop its legs over a drifting log or a natural mangrove raft, and you can see the problem.

This is why biogeographers were stunned to discover the fossils of pygmy elephants and pygmy mammoths on many different islands—including the Mediterranean islands of Sardinia, Sicily, Malta, Crete, and Cyprus; a number of islands in the Malay Archipelago, like Java, Celebes, Flores, and Timor; and the islands of Okinawa and Miyako off East Asia. Populations of mammoths have also reached the Channel Islands, off the coast of California, and Wrangel Island, in the Arctic Ocean, northeast of Siberia. This has led to speculation that many of these islands were once part of the continental landmass.

Then, in a 1980 article in *The Journal of Biogeography*, Donald Lee Johnson pointed out the surprising fact that elephants are "distance swimmers *par excellence*."[7] According to Johnson, they swim with a "lunging porpoise style" and use their trunks as snorkels.[8] More, they will take to the sea on purpose, especially when they can see a distant island and smell its tempting vegetation. This has made elephants and mammoths some of the most persistent near-island colonizers of all mammals larger than a rat.

The now-extinct pygmy mammoths of the Channel Islands off California help illustrate the differences in the two main branches of biogeography, both of which stand upon the two principle causes of organic isolation. So far, since we have given our attention to island distributions, we have focused mostly on dispersal. In Darwin's day, when the surface of the globe was believed to be much more static, this was thought to be the primary force for evolution. Plants and animals moved to different parts of the globe, ended up isolated purely by their random wanderings, then adapted to the different conditions. This view particularly gratified the minds of Victorian biologists who, even after accepting Darwin, still liked to think of a specific and localized Garden of Eden whence life ventured forth. For some species, this view is certainly accurate, but, as we shall see, for many species dispersal, in and of itself, is not an effective mechanism for isolation-induced speciation. For if certain members of a population can reach a particular area, other members of the population can reach that area too.

Thus, evolution often proceeds in the manner that Wallace and Leon Croizat suggested—with Earth and life evolving together. In these cases, species will first expand to fill out their natural range, then their distribution becomes fractionalized—sectioned off by some sort of environmental event. A valley is flooded, a glacier creeps between the population, the sea rises above or pours off an isthmus, a mountain range forms, an island sinks, a desert expands, or continents separate. The process by which a natural barrier divides the range of a population is referred to as *vicariance*. And that leads us to one of the fundamental questions that modern biogeographers attempt to address: is the distribution of closely related species the result of dispersal or vicariance? Did they cross the barrier or did it form between them?

It may at first seem that the diversification of the now-extinct pygmy mammoths or still extant dwarf foxes of the Channel Islands was the obvious result of dispersal-induced speciation events. After all, they did manage to cross the slip of seawater to the island when sea levels were at their lowest. But the isolation of these taxa was still the result of changing landscapes. The mammoths and foxes reached the Channel Islands during the glacier era of the Late Pleistocene (perhaps 20,000 years ago) when sea level was perhaps as much as 330 feet (100 m) lower than today, extending the Californian coastline and exposing all the land connecting the three major islands. The resulting "super island," called Santarosae, would have stretched to a mere 4 miles (6 km) off the coast of California—as compared to the 13-mile (21-km) distance today. Large mammoths and foxes likely swam to this massive island at this time, the former almost certainly on purpose and in search of food.

As sea levels began to rise, eventually flooding and dividing the island into three smaller isles and increasing their distance from the mainland, these mammals of the Channel Islands became permanently marooned—ensuring isolation from their Californian relatives. Since it is the rising sea level that led to their permanent isolation, this is more aptly described as vicariance.

Like the pygmy mammoths of the Channel Islands, essentially all of the fossil elephants and mammoths found on all other islands experienced significant dwarfing, and some like the dwarf elephants of Malta only reached the size of a large dog, standing just 3 feet (1 m) high. Size reduction is a well-known evolutionary consequence that usually happens whenever a larger mammal becomes isolated in places of limited resources. Other island examples include pygmy hippos, pygmy deer, pygmy foxes, and the peculiar "mouse-goat"—an extinct species of dwarf sheep that had lived on the Balearic Islands of Spain.

Larger mammals are voracious, and regions with limited resources (and no larger predators) will thus bring about their diminution. In relatively restricted places, smaller individuals, no longer persistently victimized by predation, can maintain a fulfilling and healthier diet on less food and are more resistant to starvation than their larger brethren—so the elephants, hippos, and deer that reached some of the nearby islands became miniature versions of their mainland relatives.

Reptiles and amphibians do not adhere to the same strict rules and will often experience gigantism or dwarfism on islands depending on the environment. Consider as a case in point the giant tortoises of Galápagos, or the largest living lizard, the Komodo dragon of the Komodo and Flores Islands, which can reach 10 feet (3 m) in length. Likewise, the 2-lb. (1 kg) giant chuckwalla, an iguana found on the tiny islands of the Gulf of California, is three times larger than its closest mainland relative. On the other extreme, Gardiner's Seychelles frog from the Indian Ocean island just north of Madagascar grows no larger than an ant, and many other islands have dwarf snakes.

Island isolation affects mammals differently than the more ancient classes of vertebrates because they have vastly different energy requirements. Mammals, like birds, are "warm-blooded" or endotherms (*endo* meaning inside and *therm* meaning heat). They are able to generate internally their own heat. Reptiles and amphibians, in contrast, like fish, are cold-blooded or ectotherms (*ecto* meaning "outside"). They must rely on the environment to heat or cool their bodies. Both systems have pros and cons. Mammals can maintain their own warm body temperature even at night or in the cold, but, in so doing, they must expend a significant amount of energy, which means they must be big eaters. Roughly 80% of the calories that mammals ingest are used to stoke the fires of their internal furnace.

This also means that size is very significant to mammals. Even in tropical regions, the surrounding temperature typically bathes mammals in air much cooler than their 98 °F (37 °C) body temperature. Smaller mammals find this most problematic because they have a high surface-area to volume ratio—which is to say, they have a much greater amount of exposed skin relative to their heat-producing tissues. Thus, shrew and mice have to develop extremely high metabolic rates in order to keep replenishing the heat they continuously lose to the environment—eating almost constantly. Chris Lavers, an expert on biophysics, points out that mammals could never become as small as that ant-sized frog on Seychelles—and he contends that the Etruscan shrew, at 2 g, has reached the minimum possible size for a mammal.[9] As Lavers notes, the Etruscan shrew must eat 130% of its body weight daily—compared to only 4% for the elephant. Their hearts are extremely large for their body size, more than three times larger, proportionally speaking, than that of a racehorse, and they pump at an astonishing 1200 beats per minute. Their metabolic systems are like the fiery-hot, overworked furnace of a winter house with all the windows open. They almost cannot crank out the heat fast enough.

These energy constraints explain a great number of biotic patterns. It is why the mammals of the Arctic tend to be large—like polar bears and caribou. They can retain their heat for longer periods of time and need to expend less metabolic energy relative to their size to maintain their temperature. This is also why there are no snake-shaped mammals, which would increase their exposed surface area relative to their volume to an unsustainable extent. The problem of heat loss is also why rodents, which would not find the resources of an island as limiting as an elephant, are the only mammals that tend to become much larger on islands. In a land without predators, rats are able to evolve to a size with a more relaxed metabolic rate.

Reptiles have different concerns. By relying on the environment for thermoregulation, they only require roughly 10% of the calories of mammals of similar size, but must as a consequence remain somewhat sluggish subordinates to the elements. When it is cool, reptiles cannot expend much energy and must wait for warmer temperatures to try to search for food or find a mate. This is why we do not find many reptiles in the colder climates and none in the Arctic. While on an hourly basis, hunger is the primary concern and constant motivation for mammals, the frequent worry for reptiles is temperature control.

Once again, the marine iguanas of Galápagos provide us with a perfect illustration. During the early morning, after the night has cooled them down, they act like reptilian sundials, turning their bodies at different angles to the sun in order to achieve an optimal temperature. At dawn, they align themselves broadside to the rising sun, absorbing as much heat as possible. Within an hour, when the iguanas are sufficiently warm, they point themselves directly at the sun, only exposing their chest to its rays. Soon, overheating becomes a problem, and they cram themselves into the few shaded crevices that they can find among the sun-baked lava rocks of Galápagos. At midday, when the heat starts to torment, they take to the sea and begin foraging for seaweed. Within a few minutes of their swim, their body temperature will have plummeted by 10° (5.6°, Celsius scale) or more. If they stay too long in the sea, they will not have the energy to struggle back onto the rocks. Once back on a rock, they stretch themselves out, presenting as much of their body to the warming rays of the sun as possible. Until their temperature returns to normal, they will not have the energy to digest the seaweed that sits in their stomachs. If at any point you look at a reptile in the wild—say a garden snake—and wonder what it is feeling, a good guess would be that it is feeling too hot or too cold. And accordingly, if at any point, you look at some small mammal in the wild—say a raccoon or a rabbit—and wonder what it is feeling, more than likely it is hungry.

As a result of this thermal difference between mammals and reptiles, they are often affected by island isolation in different ways. Since reptiles do not need as much food they are not impacted by limited resources to the same extent. Instead, with reptiles, the environmental circumstances of the island—particularly the dynamics of the closest niche available to them—are the primary factors that influence size. If they are carnivores themselves, they can experience gigantism as they grow into the top-predator niche, that is, if there is prey on the island large enough to drive the increase in size.

The giant Komodo dragon has evolved to become the top-marauder on its island. This 10-foot long lizard is known to hunt deer, pigs, water buffalo, horses, even humans. And their predatory method is truly fiendish—at least from a mammalian point of view. They ambush their victims and inflict a large poisonous bite that causes severe pain, stops the blood from clotting, and leads to infection. If they are not able to take control of the prey immediately, the dragons simply follow the bleeding and hopeless victim from a distance, frequently tasting the air with their long forked tongue, sampling their future meal. Up until 2005, biologists had believed that only two species of lizards, the gila monster and the Mexican beaded lizard, were venomous and that the paralysis and pain caused by a Komodo dragon bite was the result of deadly bacteria that infests its filthy mouth. But a recent article in the journal *Nature* confirmed that Komodo dragons and other members of the varanid reptile family are also venomous, and perhaps even more surprisingly, iguanas possess poisonous oral glands as well.[10] This discovery led to a mini-revolution in the theory of snake and lizard evolution. Previously, it had been believed that venom evolved in snakes about 100 million years ago, long after lizards and snakes had diverged from a common ancestor and that the two species of venomous lizards had each evolved its poisonous capability independently. The 2005 study instead indicates that all lizards and snakes are actually descended from a common venomous ancestor that lived perhaps 200 million year ago. And those snakes and lizards today that are not venomous, like, say, boa constrictors or geckos, have, at some point in their evolutionary history, lost their poisonous talents.

In 1987, Jared Diamond, the author of *Guns, Germs, and Steel*, offered an explanation for another evolutionary mystery involving the Komodo dragons. As Diamond noted in an article in the same journal, *Nature*, all the large prey that the Komodo dragon feasts upon today—the deer, pigs, and water buffalo—were introduced to the islands by humans and so have been added to the dragons' diets only recently. So if the size of this prey cannot be used to explain the lizard's gigantism, what had caused it? Diamond gave the answer in the title of his paper: "Did Komodo Dragons Evolve to Eat Pygmy Elephants?"[11] (One of the things that sparked my interest in biogeography was the opportunity to study papers with titles like that.) The fossil range of the Komodo dragons and that of pygmy elephants coincides on a few islands, particularly Flores—and invites a rather fantastic image, a lizard attacking and eating an elephant.

Flores has also now become particularly famous for another dwarf species—a diminutive group of hominids, *Homo floresiensis*, that stood only 3 feet (1 m) tall and survived on the island for some time, perhaps as late as 13,000 years ago. We will discuss them in the chapter on human biogeography, but this should give an idea of the evolutionary crucible that Flores had become. To imagine these dwarf hominids, or "hobbits" as they are sometimes called, using spears to attack pygmy elephants or to ward off the giant Komodo dragons is to imagine something out of a Tolkien novel—the Middle Earth of *Lord of the Rings* on a Southeast Asian island. But the situation is not as atypical as it may seem. Many other islands also have fauna that seem like a hodge-podge of mythical archetypes combining both the Lilliputian and Brobdingnagian—from the colossal and now-extinct elephant bird of Madagascar to the ant-sized frog of Seychelles. But what seems so alien to us continent-dwellers is not surprising at all given everything we know about island biogeography and the evolutionary forces that islands can exert on their inhabitants.

CHAPTER 4

The Volcanic Ring That Changed the World

How evolution and plate tectonics have operated in
concert to govern the major biotic patterns and
produce the Earth's most fascinating creatures
in the most exotic locales

If I had to point to a single geological feature of the Earth
that had the greatest biogeographical consequences on the
post-dinosaur world, that is, on our entire global ecosystem as
encountered today, I would choose the chain of seafloor spreading ridges that presently encircles Antarctica. This little-known
ring of cracks in the ocean bottom has drastically influenced the
geographical positioning of continents and oceans around the
Earth, which in turn, has forever altered the flora and fauna of
each of the biogeographical realms. The resulting motion of continents has galvanized biodiversity in many regions around the
globe while providing an isolating protection for certain plants
and animals in others. And this system of ridges has killed as
well—leading to the death of nearly all vertebrate life in Antarctica. Indeed, as we shall see in Chapter 7, the resulting arrangement of continents has even had an enormous effect on the
fates of human civilizations, contributing to the great differences

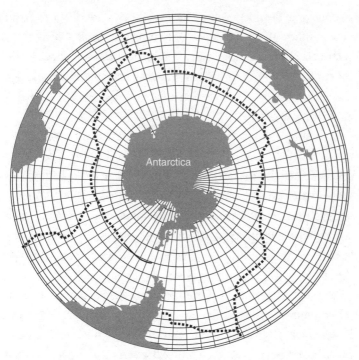

3. The rift system of volcanic seafloor spreading ridges that surrounds Antarctica has isolated it from the other southern continents and helped break up Gondwana.

in technological progress among different populations. All of this has been the by-product of the undersea volcanic ridge that currently surrounds our southernmost continent (see Figure 3). Indeed, it is the very reason that the southern hemisphere is so oceanic while the northern hemisphere is disproportionately continental.

Spreading ridges are the cracks in the surface of the Earth that snake throughout all oceans and mark the divergent boundaries between tectonic plates. That is, they are the gaps where hot magma flows up from the mantle and solidifies to form new seafloor along the edges of two plates moving away from each other. They are the mostly hidden-from-view mechanism for continental drift for which Wegener searched but could not find—the place where ocean basins are born, pushing continents apart as they grow.

Spreading ridges often begin as continental rift valleys, like that currently running through Eastern Africa from the southern tip of the Red Sea in the north to Mozambique in the south. This long fissure that is slowly carving the African continent is strewn with volcanoes, rift lakes, geysers, and hot springs—an elongated strip of geothermal Yellowstone-type features, all gurgling and boiling as the Earth's thin outer shell cracks open and exposes the effects of the hot furnace below.

Just such a rift region started splitting the Pangean continental configuration more than 150 million years ago, forming a long natural fissure along the eastern edge of the Americas and the western edge of Europe and Africa. As the supercontinent continued to fracture, the salty rift lakes along this depression started to congregate to form small seas. Eventually, as the rift continued to widen, the global seawater system poured into it. This was the birth and nonage of the Atlantic Ocean (see Color Plates 1 and 3).

The system of mid-ocean ridges that currently surrounds Antarctica helped continue the disintegration of Pangaea by breaking up its once contiguous southern block—an event that significantly transformed the history of life on Earth. Antarctica was the only Gondwanan continent that touched all the other ones—South America, Africa, Indo-Madagascar, Australia, and New Zealand. But the formation of the young ocean all around Antarctica has pushed all these tectonic plates—and the Pacific Plate as well—northward toward the equator, leaving Antarctica alone at the bottom of the world (see Color Plate 3). This has had a rather fascinating geometrical consequence. The equator is obviously the largest latitude—the east–west slice of the Earth where it is fattest, wrapping around the full circumference. None of the other latitudes describe the full circumference of the Earth. Instead, as you approach the poles, these parallel latitudinal rings become smaller. The circle of latitude at 60° South is half the size of the equator (0°). (This is not true of longitudes, which are not parallels but all meet at the north and south poles. All longitudes are the same great size—describing the full circumference of the Earth in the north–south direction just as the equator does in the east–west direction.) The result of this northward motion was that the southern continents have had to spread apart as they moved to larger latitudes, radiating away from each other (see Figure 4).

This is why the southern hemisphere is so young and oceanic. The circular bands of the Earth's crust that occupied the smaller circumferences near Antarctica have been pushed northward toward the larger latitudinal circumferences—and these southern bands of crust simply did not have enough material to occupy these regions.[1] What filled in the missing space? New seafloor crust in the Pacific, Atlantic, and Indian Oceans. Surveys of seafloor age confirm that over the past 60 million years, the most prolific ocean-floor production in the world has occurred in the southern hemisphere—particularly in the east–west direction between the Gondwanan regions (see Color Plate 3). Much of this new oceanic material was needed to fill the gaps created by the plates moving north toward the equator.

The Break-up of Gondwana

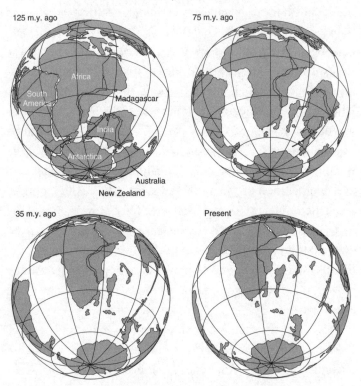

4. As the Gondwanan continents moved northward, away from Antarctica, they steadily became more dispersed along more latitudinally expansive stretches of the globe. They, in a sense, radiated outward from a center as if following the spokes of a wheel. The resulting effect of this southern hemisphere spread on all things biological—including eventually the history of human civilizations—would be difficult to overestimate.

Thus, the circum-Antarctic seafloor ridges are the reason why the bottom of the globe is so blue—the reason why the southern hemisphere has such massive Southern Pacific, Indian, and Atlantic Oceans, broken by the occasional beiges and greens of the isolated continents. It is impossible to have a full under-standing of the current organic complexion of any of these regions—Antarctica, Australia, New Zealand, Indo-Madagascar, and South America—without referring to their former Gond-wanan grouping and the resulting isolation that followed the development of the circum-Antarctic ridges. North America and Eurasia did not suffer such fractionalization, and today, they remain in essence a single great continent, connected via a now-flooded arm that stretches beneath the Bering Sea and Strait. But what had once been a single landmass below the equator, a tight cluster of continents that shared many plants and ani-mals, has now divided into many diverse and stunning realms. Indeed, Antarctica, Madagascar, Australia, and New Zealand have all become islands—and for a while—South America was as well.

Perhaps, the most obvious biological consequence of the devel-opment of the circum-Antarctic ridge is also the most disturbing. As the surrounding ridge systems steadily isolated Antarctica, all of its non-flying and non-marine animals became trapped on a continent that was about to be overtaken by one of the most destructive forces in Earth history: ice.

Many people tend to think that the current state of the poles, with Antarctica smothered beneath 1.9 miles (3 km) of glaciers, as "normal," that the way the world is today is the way it is supposed to be and must remain. The truth is that for much of the past 250 million years, the Earth was usually much hotter than it is now—often with little to no year-round ice anywhere. During one of Earth's warmest points, around 50 million years ago, London had an average temperature of 77 °F (25 °C), and it had tropical forests—as did Montana. The very high latitudes during this time have been described by the geologist Donald R. Prothero as the "the palmy balmy polar regions,"[2] and Ellesmere Island in the northernmost section of Canada, deep within the Arctic Circle and adjacent to northern Greenland, was home to turtles, salamanders, and alligators. Ellesmere had a climate similar to what we might find in the middle of the United States today. Antarctica, too, was a thriving, temperate forested ecosystem—even though it was pretty much in the same place where it is now. The fauna of the Antarctic Peninsula of the mid-Eocene (perhaps around 40 million years ago) was very much like that of southern South America at the time, with a great number of birds, including large, flightless ratites, perhaps similar to the rheas of South America today, and penguin ancestors, which were quite comfortable in the mild weather, having not yet adapted to the brutal cold. A healthy variety of mammals also roamed these Antarctic forests, including large hoofed animals, sloths, and numerous kinds of marsupials (Figure 5).[3]

Since climate is such a chaotic system consisting of innumerable variables, we do not yet understand why the globe started cooling so significantly at the end of the Middle Eocene (~37 million years ago) and, again, at the "terminal Eocene event" (~34 million years ago), when the world abruptly went from greenhouse to icehouse. Many possible causes including declining levels of carbon dioxide have been suggested. But climatologists believe that one of the most significant factors was the final separation of Australia—and specifically the South Tasman Rise—from Antarctica. It was then and there that the seafloor spreading ridges completed the circle around the southern continent, allowing the development of the circum-Antarctic current system and greatly decreasing oceanic transport of warm

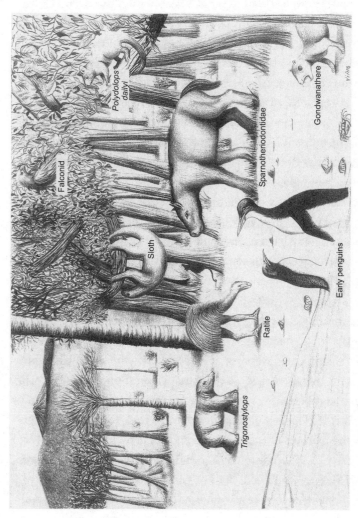

5. Antarctica, around 40 million years ago, not long before the ice.

waters to the southern continent. Surface water temperatures around Antarctica dropped by as much as 14–15 °F (8 °C).

And that was when the glaciers started to come.

In terms of number of species lost, the Earth has faced far more devastating crises—particularly the global extinction events that marked the end of the Permian and Cretaceous eras. But on a continent-wide level, in terms of *percentage* of species killed and never replaced, perhaps nothing can compare to the Antarctic catastrophe—not meteors, not supervolcanoes, not the evolution of humans. As far as we are aware, at a continental level, nothing has been so efficiently destructive, so indiscriminately murderous, as ice.

After hundreds of millions of years of liveliness and greenery, Antarctica in a very brief period at the end of the Eocene became an icy, barren wasteland. Antarctica is much colder than the Arctic because the latter sits atop an ocean—a massive and dense heating system that remains near or above 32 °F (0 °C). Antarctica, by contrast, is a frozen block of land, far too cold to support the relatively diverse mammals that inhabit the icy north. Today, the temperatures of Antarctica in the winter are similar to the surface of Mars—and it seems every bit as inhospitable. It is the highest, driest, coldest, and windiest of continents. And what complex life it does support, the few plants and simple insects like springtails and mites, must cling to the ice-free tips on the margins of the continent and some rocky regions in East Antarctica. The continent is roughly the size of United States and Mexico combined, but less than 1% of it is currently available for plants to take a hold. It has no trees or shrubs and only two flowering plants: pearlwort and hair grass. The other plants are typically the ground-hugging ones that are best equipped for extreme environments, like mosses and lichens. There are some ice-free regions in different parts of the Antarctic interior, where the dry conditions and lack of snowfall prevent glaciation. But these so-called "dry valleys" harbor almost no terrestrial creatures at all except for the little roundworm nematodes, which feed off bacteria and yeast. The food chain of these barren regions is, as far as we can ascertain, the simplest on the planet, and nowhere else is the soil so poor in ecological diversity.

While Antarctica may often be striking to look at in pictures, when the summer sun brightens the ice, we should not forget that beneath the ice sheets is a continent-wide boneyard. The notorious end-Cretaceous extinction event that wiped out the dinosaurs, though it did occur globally, did not have as devastating an effect on any particular continent as did the slow spread of glaciers on Antarctica. At the end of the era of dinosaurs, roughly 65–85% of all species perished, but the suppression of biodiversity did not last long, and the gain in mammal and bird species helped offset the loss of dinosaur species. But this was not true in Antarctica. Its isolation and freezing simply brought to that continent an endless death—taking more than 99% of all its species. One understands the fascination and reverence that geologists and climatologists have for the austerity and blistering ruggedness of Antarctica, but I suspect that not a few biologists sympathize more with the view of Robert Falcon Scott, the British explorer who reached the South Pole in 1912. "Great God, this is an awful place," he wrote just before he succumbed to cold and starvation, along with the rest of his team.

One also imagines the confusion that must result when biologists watch recent glorifications of the current state of Antarctica by people expressing their understandable concerns over global warming. Certainly, the argument that humans should strive to reduce their impact on the world is sensible, but we should be careful not to make glaciers seem like planetary life-preservers or the ever-vigilant custodians of biodiversity. This common misconception is probably not helped by the various slide shows and films on global warming that include pictorial ballads to glaciers that bemoan evidence of their melting with plaintive music. While the point being made is well taken, this still could seem to some biologists as quite as misguided as creating monuments to malaria or odes to desertification. No matter how much we want to preserve them, we still should not forget that glaciers are, quite simply, crawling death, the most efficient long-term suppressers of biodiversity of which we are aware, and a steady and unstoppable force that destroys everything green.

A more realistic view of the ice caps now smothering Antarctica also helps us to reassess our perception of its most famous denizen—the emperor penguins. It is tempting to look at these swimming birds as little tuxedoed, half-ridiculous waddlers, happily frolicking in a winter wonderland. But their cute awkwardness on land not only gives their homeland a misleadingly benign appearance, it overshadows the penguins' most significant attribute—their almost inconceivable tenacity. The truth is that emperor penguins provide the single most gripping example of rugged survivalism, perhaps in the entire history of backboned creatures. They are the only remaining holdouts of the diverse year-round vertebrate fauna of the Antarctic Eocene. As all the other native creatures continued to fall before the advance of the glaciers, as all of the Antarctic placentals, marsupials, reptiles, birds, and freshwater fish went extinct, the ancestors to the emperor penguins continued to adapt and persist. In contrast to the Disney-like view of the emperor penguins, we should see them for what they really are: the stressed and brutalized victors of the largest and most vicious game of survivor in the past 60 million years. In the dead of winter, on the entire mainland continent of Antarctica, they are now all alone—the only creature left standing.

The keys to their survival included both physical adaptations—like their layer of blubber, dense coat of feathers, and large size, all of which help to retain heat—and an array of complex sociobiological innovations. The old saying that "it takes a village to raise a child" is really true for the penguins, which rely on the entire colony in an effort to get their eggs and chicks through the winters. As the cold season approaches, the penguins travel away from the sea and deeper into Antarctica to breed and lay their eggs—sometimes as far away as 125 miles (200 km). They have to reach ice that will not break apart beneath their chicks during the warmer season before they are ready to swim. In May, after each mother has laid a single egg, they transfer it quickly over to the father, who carries the egg on his feet and protects it from the cold with a fold of skin on his belly. For two mostly dark months, the fathers endure the Antarctic winter incubating the egg, while the mothers, who have invested so much energy in producing the eggs, return to the sea to fatten up on fish, krill, and squid.[4]

The emperor penguins have developed a close herding instinct—and are the only penguin species that is not territorial. During the brutal winter blizzards, when temperatures can drop to −40 °F (which is also −40 °C), the father penguins, eggs on feet, will huddle together for warmth, sometimes for days. The penguins steadily shift positions during their huddling, so that no penguin is exposed for too long on the outside of the group. The eggs hatch in mid-July, right about the time the mothers return with food in their crop for the new chick. Counting the original trek and courtship, the male emperors have now lasted for four months without food and have lost half their body weight. They soon head back to the sea, refuel, and return again to help tend the chicks. The mother and father thus take turns, trudging back and forth from the sea to the chick, helping the young penguin grow and build strength. We often look at various behavioral adaptations as merely helpful nudges, as a trait that helps bestow a selective "advantage" on some individuals and so starts to become widespread through the population. But with the emperor penguins, this complicated suite of urges and instincts was indispensable to their survival. Without the development of the intense sacrifice of both parents and the full cooperation of the huddling penguin colonies, they almost certainly would have become extinct—just as everything else around them did.

So while we tend to think of penguins as naturally cold-weather creatures, the truth of the matter is that they were forced to endure the advance of the ice—and the process was merciless.

The biological effect on the other continental regions that continued to inch away from Antarctica, though not as ruinous, was still certainly drastic. As the seafloor continued to form around Antarctica, the rest of the Gondwanan continents continued to move toward the more latitudinally expansive parts of the world, leading, in most cases, to the isolation-induced peculiarity of their flora and fauna. In other words, these northward jaunts to loneliness are why the plants and animals of Australia, New Zealand, and even South America seem so exotic.

Ironically, the final separation of Australia and Antarctica—the very thing that doomed almost all life on Antarctica—is what helped save the marsupials of Australia. The rats and dingoes are relatively recent arrivals, and so for much of the past 40 million years, the pouched mammals of the island continent never had to compete against the most successful of the placentals that evolved after the Eocene and which helped eradicate nearly all other marsupials in other parts of the world.

One of the more conspicuous engines for biodiversity is continental breakup, which must produce differentiations among all the creatures and plants previously common to both regions. The separation of South America from Africa cleaved all those species that the continents shared into South American and African varieties. Birds, mammals, freshwater fish, spiders, insects—indeed all taxa that lived in both regions had to differentiate into two main groupings. Perhaps most significantly, the start of the rifting between Africa and South America a little more than 100 million years ago appears to coincide with the division of two of the four major groups of placentals, the Afrotheria (e.g., elephants, aardvarks, manatees, elephant shrews) from the Xenarthra (e.g., anteaters, armadillos, sloths).

Afrotheria, which as the name suggests evolved from an African ancestor, is only a recently recognized (and still debated) grouping of mammals that includes some of the largest (elephants) and smallest of the mammals (elephant shrews). This new ordering came about through careful analyses of DNA and protein sequence data—and was certainly quite surprising.[5] Prior to the molecular analyses biologists believed that elephant shrews were, well, shrews—an African version of the European mouse-like creature. The long nose, which the elephant shrew can twist about in search of insects, spiders, and worms, was thought to be the result of adaptation to local conditions—a chance superficial resemblance to the trunks of the elephants that roamed nearby. In reality, its nose was the key biological hint, and it was the "shrew" part of the name that was misguided. Elephant shrews have a much closer relationship to elephants than they do to shrews.

Other members of the Afrotheria also have sensitive and versatile noses. The aquatic dugongs and manatees have large muscular snouts that help them dine on seagrass and other forms of aquatic vegetation. And the nose of the aardvark is also long and tactile. Note too that the anteater of the South American Xenarthra also has an extremely pronounced sniffer. This perhaps suggests that the original ancestor of the Afrotheria and Xenarthra that roamed western Gondwana may have been an insectivore with a long nose. Then, after Africa and South America began to rip apart, dividing the populations of this fruitful ancestor, the African group developed an even more sensitive and flexible proboscis. Over the course of the next 100 million years, many of its descendants would continue to rely on and employ their rather useful snouts. Thus, the nosey Afrotheria almost certainly came from an original nosey ancestor, and these lines of descent eventually gave rise to the most famous proboscis of them all, the elephant's trunk.

A classic example of a Gondwanan distribution involves the ratites, the typically huge, flightless, fast-running birds that include ostriches and their closest allies. Essentially, each of the isolated southern realms boasts its own example of a ratite—rheas (South America), ostriches (Africa—and fossils in Eurasia and India), the extinct elephant bird (Madagascar), the emu (Australia), the extinct moa (New Zealand), and the chicken-sized kiwi (also from New Zealand). Fossil sites in Antarctica have also confirmed that even the now-frozen continent had once been home to a ratite. Recently, researchers determined that the tinamous, a chunky partridge-like bird from Central and South America, should also be included within the ratite clan. This surprising find has challenged traditional theories of how and when ratites became flightless because the ground-dwelling tinamous is still able to fly, albeit weakly.[6]

Of all the ratites, the fossils of the Madagascan elephant bird deserve special attention. Standing over 10 feet (3 m) tall and weighing nearly half a ton, this feathered colossus laid the largest bird egg known—roughly 3 feet (1 m) in circumference and holding nearly 2 gallons or roughly the same amount of volume produced by 150–200 chicken eggs. It is possible the elephant bird lived as late as the seventeenth century, and many suspect that it was the source of the Arabian legend of the "roc," the mythical avian terror said to have carried off young elephants in its beak and attacked the ship of Sinbad.

The fragmented distribution of these typically flightless giants would seem to indicate that their most recent common ancestor had been widespread about Gondwana. And a molecular analysis by an Oxford research team headed by Alan Cooper has supported this supposition, suggesting that the ratites on the different plates all diverged over a 20 million-year period in the Late Cretaceous (i.e. from ~90 to 68 million years ago), which is consistent with the Gondwanan disintegration. Much later, around 35 million years ago, the speciation of the Australasian ratites, the emus and the cassowaries, then followed. Cooper and colleagues described their comparison of the ratite genetics with plate tectonics as the "first molecular view of the break-up of Gondwana."[7]

The ratite diversification, of course, did not precisely match the appearance of each mid-ocean ridge. But that is to be expected. At times, the earlier formation of rift valleys, with their deep lakes and volcanoes, may lead to the division of certain populations of plants and animals long before seafloor begins to form between the regions and continental breakup has officially begun. Still other creatures particularly amenable to crossing narrow marine barriers may not become genetically isolated from their relatives on the departing continent until it has become relatively distant. Cooper and colleagues suggest Antarctica was a central conduit for ratites, and their chronology of diversification does seem most consistent with Gondwanan continents breaking from the southern continent—although it is possible that the recent inclusion of tinamous within the ratite group may alter this biogeographical scenario.

The following map (Figure 6) and evolutionary tree, referred to as a *cladogram* (Figure 7), describe both the distribution and evolutionary history of a tiny and colorful freshwater killifish group known as the Aplocheiloidei. As we shall see, the evolutionary history of the killifish precisely coincides with the geological history of the regions they inhabit.

As the cladogram shows, the Gondwanan killifish has steadily diversified into a greater number of groups (families, genera, species) as Gondwana continued to break up. First, a population of the ancestral killifish inhabited a large section of Gondwana, and part of this group became isolated on Indo-Madagascar. The extreme age of the seafloor between Madagascar and Africa suggests the great continental island had separated from the continent as much as 135 million years ago.

Next came the division of Africa from South America, which did not occur until 100 million years ago or perhaps even later. In 1981, Lynne Parenti, curator of fishes at the Smithsonian and a renowned biogeographer, was the first to divide the African

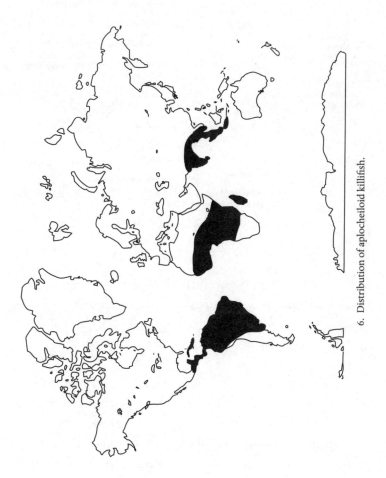

6. Distribution of aplocheiloid killifish.

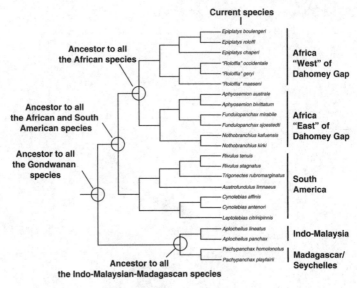

7. Cladogram of aplocheiloid killifish showing evolutionary history.

and South American Aplocheiloidei into two closely related but distinct families, and she recognized this precise cleavage as indicative of the African–South American geological split. The analysis by Murphy and Collier supports her biogeographical analysis.

The Seychelles are islands comprising a sliver of a Gondwanan continental block just north of Madagascar and formerly connected to it. When India finally separated from these Indian Ocean continental fragments, this divided another group of Aplocheiloidei into Indo-Malaysian and Madagascan-Seychellean forms. Also, in Africa, the development of intra-continental barriers—perhaps due to regional flooding and the eventual formation of a dry savanna region known as "the

PLATE 1: *The History of the Atlantic Ocean.* We can trace the formation of the Atlantic Ocean by studying the age of its seafloor. (See Color Key with Plate 3.) The rift system (thick black line) that originally carved the matching outlines of the Americas, Eurasia and Africa currently runs through the middle of the Atlantic Ocean. This long volcanic fissure is where new seafloor still forms today.

PLATE 2: *The History of the Indian Ocean.* As in the Atlantic, the age of the seafloor has faithfully recoded the development of the Indian Ocean. The evolutionary histories of many plants and animals parallel the timing and formation of the Atlantic and Indian Oceans..

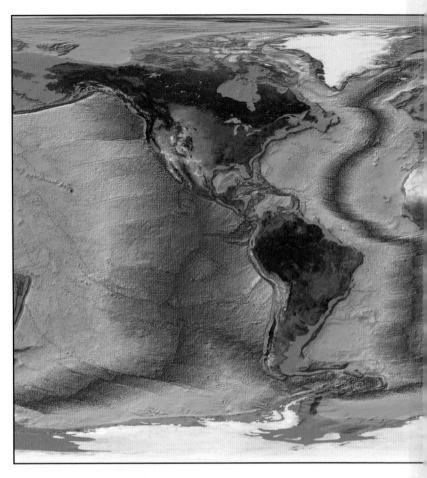

Age of Oceanic Lithosphere

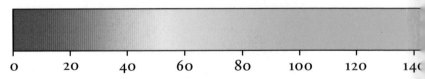

| 0 | 20 | 40 | 60 | 80 | 100 | 120 | 140 |

PLATE 3: Crustal Ages of the World's Seafloor. The chronological constitution of oceanic crust, as indicated by the Color Key, provides an extremely useful overview of the history of ocean basins, often showing precisely when and how continents have split apart. The resulting continental arrangements have shaped the biogeographical patterns observed today, preserving ancient, exotic, and vulnerable plants and animals in the south while contributing to the rise and domination of northern mammals. (Crustal age figures provided by image author, Elliot Lim.)

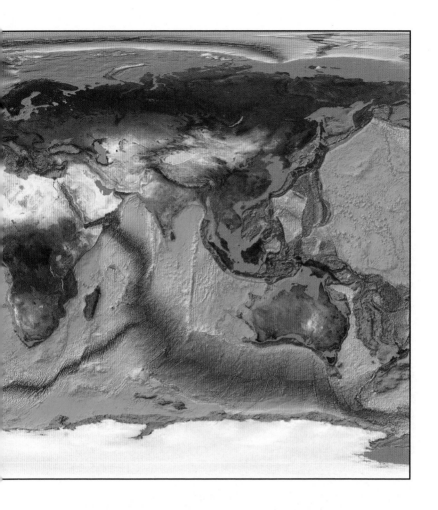

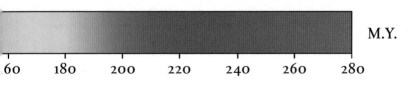

M.Y.

60   180   200   220   240   260   280

PLATE 4: The giant ground sloth, *Megatherium*. Darwin's discovery of a nearly complete fossil skeleton of the tree-pushing *Megatherium* in South America—where all modern species of sloths are still restricted today—reinforced that same biogeographical bond that he found with so many other plants and animals.

Dahomey gap" that split the African rainforest into eastern and western sections—resulted in a purely African partition among these killifish.

The evolution of killifish also helps instruct us on various terms and concepts used in biogeography. First, note that the Aplocheiloidei of each region are *monophyletic*, which is to say, they form a complete genetic unit that includes *all* the descendants of a single ancestral population. Another way of saying this is that all the current Aplocheiloidei in a region are more closely related to each other than to any other fish outside that region. As a case in point, after Africa separated from South America, none of the African Aplocheiloidei managed to leave the continent—while none of the other Aplocheiloidei, whether in South America or Indo-Madagascar, managed to enter Africa. Had, say, any of the killifish from Africa crossed the Mozambique channel to Madagascar, then the African Aplocheiloidei would be *paraphyletic*, which is to say, some of the killifish in Africa would be at least as closely related (indeed, in this case, more closely related) to those recent migrants in Madagascar than to other African Aplocheiloidei. The easiest way to think of the difference between monophyly and paraphyly is that a house with a woman and her only three daughters is monophyletic. However, if one of the daughters moves out or if one of their cousins comes to visit, the household group becomes paraphyletic, for the household no longer forms a complete genetic unit. Thus, "dinosaurs," as classically defined, comprise a paraphyletic group. The traditional categorization of dinosaurs does not include birds and so does not comprise all the descendants of dinosaurs. Monophyletic groups (also sometimes known as "clades") can come in any size. The woman and her three daughters form a very small monophyletic group; mammals form a larger monophyletic group; and indeed all life on Earth, including all ancestors, comprise the largest monophyletic group.

The specific branch of evolutionary science that focuses on determining the order and hierarchy of descent and displaying it by such symbolic trees is referred to as *cladistics*, originally developed in its modern form by the German entomologist Willi Hennig in the 1950s. Hennig's work helped advance biogeographical studies because it allowed one to use biological relationships to uncover or test geological history.

The cladogram in Figure 7 shows the relationship of just one group of fish, but many other taxa flaunt a similar geological relationship. Cichlids, another freshwater fish, also have a similar distribution, occurring in Southern India, Madagascar, Africa, South America, and the American tropics. As with the Aplocheiloidei, the cichlids of Madagascar are more closely related to those from India than they are to the African (or South American) cichlids. If we were of the view that continents never moved and only had the map of the globe that we have today, which indeed is what pre-drift geologists had to use, this would have seemed especially odd given that only the Mozambique channel separates Madagascar from Africa—while India is on the other side of a significant stretch of Indian Ocean. But the crustal age of seafloor helps explain the distribution. The seafloor between India and Madagascar is much younger than that between Madagascar and Africa, suggesting the continental island had pulled away from Africa before it had split from India. The age of the seafloor recounts the breakup of Gondwana, which in turn mirrors the division of a number of its plants and animals (see Color Plate 2).

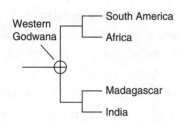

8. This areagram is consistent with the evolutionary history of a number of plants and animals.

Many other Gondwanan holdovers on Madagascar also show closer relationships to Indian taxa rather than African ones. Given enough repetition, we can look at this as a pattern that reflects the geological history of the region, allowing us to substitute the geographical regions for their resident plants and/or animals on the cladogram (Figure 8).

In other words, the recurrence of this biogeographical pattern suggests the regions may have fragmented in the same order as have their plants and animals. And the variations in the age of the seafloor between these regions support this view (see Color Plate 2).

This method, in which we search for shared evolutionary histories and use such patterns to create area cladograms, is known as *cladistic biogeography*—a process developed by Gareth Nelson and Norman Platnick in the 1970s (see also the splendid book *Cladistic Biogeography* by Christopher Humphries and Lynne R. Parenti).[8] Such biogeographical practices unite the Darwinian branches of relationships with geological history, helping us advance conceptually from the old metaphor of a stand-alone evolutionary tree, seemingly removed from its background environment, to the more comprehensive analogy of a global creeping ivy spreading across a particular landscape.

The Madagascar–India relationship is no longer that surprising, as it fits with classical reconstructions of Gondwana as well as seafloor age data. But another biogeographical pattern involving these southern realms has suggested a geological relationship that was not precisely expected. A number of taxa—including iguanids, boids (e.g. relatives of boa constrictors), turtles, crocodyliforms (e.g. ancestors of crocodiles), and gondwanathere mammals—seem to suggest links between South America and Madagascar/India, even to the exclusion of Africa. For some of these creatures, some researchers have suggested long-distance cross-ocean dispersal. But as the biogeographers John Sparks and Wm. Leo Smith observed, this would "assume that boas, iguanas, and pelomedusoid turtles not only independently dispersed across thousands of kilometers of ocean (from South America to Madagascar or vice versa), but that all of these lineages somehow also entirely missed Africa in the process!"[9]

Yet if you look at the global map of Gondwanan breakup (Figure 4), particularly the continental arrangements 125 million years ago and 75 million years ago, we see the possible explanation. First notice how close the bottom peninsula of South America (known as the Falkland Plateau) is to Madagascar at 125 million years ago. Africa was the first to separate from these southern regions, and continental rifting prior to that may have formed natural barriers between South America and Africa consisting of volcanoes and lakes. Thus, it is possible that this southern stretch of South America was in contact with Madagascar—or nearly so—when Africa had already separated from South America by the prohibitive rifting landscape or a narrow sea. Also, long after southern Africa had become separated from the other Gondwanan regions by a significant marine gap, South America remained attached to the Antarctic Peninsula, and some researchers suggest that Madagascar also remained connected to Antarctica through the Kerguelen Plateau into the Late

Cretaceous (see Figure 4 and the Gondwanan arrangement from 75 million years ago). These also may have provided corridors for the movement of land-based creatures between these regions that would exclude Africa.

As we shall see in the next chapter, it is not merely that the southern continents split apart, it is the duration of the resulting isolation that has produced some of the most well-known and striking biogeographical designs that currently paint the Earth.

The Bloody Fall of South America and the Last of the Triassic Beak-Headed Reptiles

South America, Australia, and New Zealand as sanctuaries for "living fossils"

I magine New Zealand had remained completely undiscovered by scientists until today, and that the first person to step onto its shoreline was a biogeographer. In the beginning, she would probably not suspect a thing. New Zealand is, of course, an island in the middle of an ocean. It is 1250 miles (2000 km) from Australia, and the life that clings to it, on first inspection, certainly seems to reflect its location. As with essentially all the other 25,000 islands in the Pacific, the intrepid voyagers—the ferns, terns, and other ocean island taxa—are prevalent. So it is likely that, in the beginning, nothing would seem out of the ordinary.

Eventually, our biographer would notice the kiwi, but that would not immediately elicit much surprise. Islands often have flightless birds because, with no predators and so little competition for easily accessible insects, nuts, or fruits, flight becomes an unnecessary energy drain. The flightless dodo of Mauritius in the Indian Ocean is probably the most well-known example. These massive and notoriously homely birds actually evolved from a flock of pigeons that flew there from Southeast Asia. Likewise, when the Polynesians first reached Hawaii, they found geese, ibises, and rails—all with small and useless wings. Other flightless birds of New Zealand include the world's heaviest parrot, the Kakapo, and the dark, hen-sized, Takahe, which is part of the rail family.

This tendency toward flightlessness on remote islands helps underscore what a huge energy commitment flying requires. It also highlights the current and continued ferocity of the forces of natural selection that work on the mainland organisms. For many species, islands greatly reduce two of the key pruning agents of evolution—predators and severe competition for food—so island relatives help depict what mainland birds would become like if these same cushy conditions existed on continents. In other words, the same flightless side-branches that developed and eventually took over the bird populations on the islands (weaker chest muscles, smaller wings) are also continuously being started among the continental populations. It is just that on the mainland, these side-branches are just as continuously being lopped off—often by carnivores like coyotes or cats. Once again, we see natural selection as an indefatigable sculptor.

But the flightless kiwi is not like other island birds, and if the biogeographer inspected it a little more closely, she may have realized that it was actually a ratite, a smaller relative of other larger southern hemisphere ostrich-like birds. The ancestor to the kiwi lost its ability to fly tens of millions of years ago, and the global distribution of other ratites may have led the researcher to suspect that New Zealand was not just another young oceanic island. And once she happened across the lizard-like tuatara, she would have no doubt. New Zealand is ancient and was not always so isolated.

The two tuatara species of New Zealand superficially resemble sturdy iguanas, with crests running from their heads to their tails. But taxonomists have determined that they are the last remaining species of the order Rhynchocephalia ("beak-headed reptiles"), a group that was prevalent in the Triassic more than 220 million years ago, before the rise of the dinosaurs. Currently, only four orders of reptiles survive—Testudines or Chelonia, which consists of all the species of turtles and tortoises; Squamata—all the snakes and lizards; Crocodilia—all the 'gators and crocs; and the Rhynchocephalia—now represented by the two lonely lizard-like species on New Zealand. The tuataras' closest known relative is a fossil sphenodontian of Late Cretaceous South America, which went extinct at about the same time as the dinosaurs. New Zealand has turtles as well, but no native snakes or crocodiles.

As expected, a number of the features of the tuatara are primitive and do not seem quite as efficient and streamlined as those of their more successful reptilian counterparts. Unlike modern reptiles, the males do not have a penis, but deposit sperm directly from their one opening—the cloaca—directly into the cloaca of the female, somewhat like certain amphibians do. This reproductive method seems to be on the mid-point between the evolution of internal fertilization—which is to say, sex as we mammals understand it—and the external fertilization typical of most fish and amphibians. Such aquatic taxa lay eggs in the water that are then fertilized by the free-swimming sperm recently secreted by some males. The sperm are able to find the right eggs through chemical attraction, but, this, of course, is a risky endeavor. Currents, predators, and the sperm of other possible mates can interfere. Some amphibians and vertebrates of the sea increase their chances of successful fertilization by clinging onto each other as the females lay eggs and the male secretes sperm—ensuring that both occur at the same time and same place. It is rather easy to imagine how this would lead to a more direct form of fertilization, where the male and female press their cloacae together. Such a process allowed reptiles to live all their life on land without having to return to ponds, streams, or rivers to breed.

Yet the more advanced forms of sex, even in mammals, still retain many of the same fish-like characteristics. The egg still stays stationary, and the much smaller sperm must swim to find it. It is just that with reptiles and mammals the process occurs in a safer, smaller, enclosed environment. Once again, we see how evolution occurs with very slight, incremental changes.

The teeth of the tuatara are also peculiar. They are fused into the jaw with little chance of replacement. And their eggs can take as long as fifteen months to hatch—a length of time that would make them especially susceptible to mammals. These primitive characteristics of the tuatara, which seem to accent their vulnerability, have managed to survive only because New Zealand's motion away from Australia and Antarctica has kept them protected from the more efficient vertebrates that fought and prospered amid the extreme and constant competition of the continents.

Other endemic fauna also make it clear that New Zealand could not be a young oceanic island but is really an ancient and rifted continental fragment. These include primitive frogs of the genus *Leiopelma*—which are very similar to fossil frogs of the Jurassic. They do not croak—and lack both vocal sacks and ear drums. They also go through their tadpole phase in the eggs, and then hatch as froglets.[1] This perhaps suggests that the free swimming tadpole phase of frogs represents an evolutionary advance, presumably because it is harder to catch a tadpole than an egg.

In brief, we may look at New Zealand as an ocean island treasure chest, overgrown with Polynesian plants and birds, but also harboring a number of ancient Gondwanan jewels. Conservationists are now trying to protect their unique native flora and fauna, but they must act quickly. Both tuatara and the native frogs have suffered widespread extinction on both main islands, and the tuatara and some of the frogs now only survive on tiny, unpopulated rat-free islands in between the North and South Islands or just offshore.

Not too far to the north of New Zealand is another ancient continental fragment, New Caledonia, which, like its southern neighbor, has also been pushed around by ocean formation and left in the West Pacific. Deep in the misty cloud forests of New Caledonia, we find a relict that has very recently shocked the world of evolutionary biology. A number of genetic analyses have determined that the New Caledonian *Amborella* shrub, with its greenish-yellow flowers and small red fruits, dates back more than 130 million years and is actually sister to all other flowering plants in the world. In other words, all other flowers (angiosperms)—all 250,000 to 400,000 species—from lilacs and roses to apple trees and tomato plants belong to one group— while the little *Amborella* belongs to another. Scientists have now begun studying this non-descript shrub as if they had just found the botanical Rosetta stone, and one team of scientists has called for the complete sequencing of its genome.

Unlike all other angiosperms, *Amborella* has no vessels for sucking water from the ground. They are also unisexual, either male or female, and require cross pollination before they can produce their tiny seed and fruit. Perhaps, more importantly, the embryo sac of *Amborella*, the structure that holds the egg, is different from all other angiosperms. These reproductive variations are significant because one of the more obvious differences between the two groups of plants that dominate the world—the truly ancient gymnosperms (meaning "naked seed") and the more modern angiosperms (meaning "vessel seed") involves reproduction. Conifers like pine trees are gymnosperms, and pine cones hold the uncovered seeds of their gymnosperm parent within their woody scales. Angiosperms, by contrast, encase their seeds in a fruit, which is a very helpful adaptation. Fruit not only helps the seed when it rots and juices the surrounding soil, thus providing a nutritious nursery for the young plant; fruit has also become a great transportation aid by becoming tasty to mobile birds and animals that can carry the seeds great distances before excreting them. Once again, the seed has the advantage of growing in a rich and fertilized environment. This is one of the reasons why the angiosperms have become so widespread. Thus, just as the tuatara gives us insight into those first post-amphibian ancestors of reptiles, the *Amborella* may help reveal some of the traits of that first post-gymnosperm plant whose descendants have helped paint and vivify the world.

We also find obliging clues to the evolutionary development of mammals in other Gondwanan sanctuaries—specifically in Australia and New Guinea, where live the last holdouts of the egg-laying monotremes. Currently, the world is populated by three known classes of mammals, the monotremes (the duck-billed platypus of Australia and the spiny echidnas of Australia and New Guinea), the pouched marsupials (like the kangaroos and koalas of Australia), and the placental mammals (such as panda bears, arctic hares, and investment bankers.) Roughly 250 million years ago, long before the rise of dinosaurs, the evolutionary track of mammals irrevocably veered from the reptile line. This is when we see the evolution of synapsids, which are the first mammal-like reptiles. Eventually, some of the descendants of these early synapsids branched into the three forms of mammals today—with the placentals appearing to be the farthest removed from their reptilian ancestors.

In contrast to the cozy and protective gestation of placentals, the marsupial birthing method seems like a somewhat shaky and makeshift substitution for egg-laying. They deliver their young live, but only after very short gestation periods, and in a very fragile and underdeveloped state. Their placenta is really more of a yolk sac. And, at birth, the baby kangaroo is no larger than a peanut—a blind, pink, hairless fetus-without-a-womb that must crawl on its own through the mother's fur into the pouch. The young then matures in this maternal pocket, suckling a teat for milk. This process, of course, was more advantageous for the distant ancestors of marsupials than just leaving eggs to the mercy of the elements and predators, but it still seems far dicier than the long-lasting protection offered by their unpouched cousins.

Monotremes seem to be more primitive still, looking a little like something you would get if a mad geneticist tried to create a placental–reptile hybrid. Like reptiles, they still lay eggs—and like reptiles they have one opening in the posterior for, well, everything—both reproduction and elimination of all waste. "Monotreme" is Greek for *one hole*. What mammalian characteristics they do have seem to be basic and incomplete. First, they are not quite as warm blooded. The temperature of placental mammals runs from 96.8 to 102.2 °F (36–39 °C), whereas the platypus's body temperature averages 88 °F (30 °C), and like cold-blooded animals, this temperature fluctuates. Monotremes do suckle their young, but not from teats. Nursing mothers sweat milk along their bellies, and the young suck it up from their tufts of fur. Recently, a team of evolutionary geneticists headed by David Brawand has concluded that the common ancestor of monotremes, marsupials, and placentals also lactated and that it was this evolutionary adaptation that helped promote the switch from egg-laying to live birth.[2] In other words, the mammal line shows a steady switch from yolk to milk as the primary nutrient for offspring. Thus, on the evolutionary trek toward warm blood-edness, lactation, and live births, the platypus still seems stuck in the middle.

This leads to an interesting etymological side note. The platypus was not brought to the attention of Europeans until 1798, when the body of one was sent to England (and many people believed it to be a crude hoax, a duck's bill sown on some mammal). Forty years earlier Carl Linnaeus, the father of taxonomy, had coined the word "mammal" based on what he believed to be an exclusive feature shared by all mammals—the "mamma" or breasts. However, had the teat-less, milk-sweating platypus been discovered by Europeans just a little earlier, Linnaeus would have realized that not all mammals have "mamma," though they do all produce milk. So it is possible that we would now be referring to ourselves and all our furry relatives as "lactals" rather than "mammals."

In the previous chapter, we discussed the split of two of the four groups of mammals—the South American Xenarthra and the African Afrotheria. Similarly, on the northern hemisphere continents (Laurasia), the populations of another incredibly successful ancestor also became divided, leading to the diversification of the other two placental clades, Laurasiatheria and

Euarchontoglires. Laurasiatheria, which have their place of origination in their name, is a multifaceted group that includes carnivora (cats, dogs, bears, etc.), cetartiodactyla (pigs, deer, hippos, whales, etc.), and other orders that include horses, bats, and shrews. The Euarchontoglires encompass rabbits, rodents, primates, tree shrews, and flying lemurs. It is possible that populations of the ancestor to these two fundamental clades became divided by the opening of the Atlantic Ocean between Europe and North America—just as their southern counterparts became divided by the southern Atlantic Ocean. But we do not yet know for sure.

There is still debate about whether the reproductive differences between placentals and marsupials explain the disparity in success between the two furry groups. The problem is that placentals did indeed exist in Australia (at least 55 million years ago), and they disappeared while the marsupials prospered. South America was also a region where marsupials diversified into many different niches, but they lived alongside placentals—the only region that, for a while, ended in a tie. Today, it seems that placentals dominate everywhere they have been able to get to. But actually, as we shall see, it would be more accurate to say that *northern* placentals now rule the mammalian world.

One of the main reasons for this hemispheric bias in mammalian success is that Eurasia remained linked to North America via the Bering Bridge throughout the Cenozoic—and was connected to Africa as well. The resulting massive terrestrial platform promoted the interaction and competition among a far greater number of individuals and species—a tricontinental evolutionary arms race among animals that promoted speed and guile and countless other helpful adaptations. The herbivores had to develop new talents and abilities to out-compete other plant-eaters and evade all the carnivores that were raiding their realm from every direction. This, in turn, pushed the evolution of the predators.

Each of the southern continents had far fewer individuals playing the genetic lottery, so they had fewer opportunities for their mammals to happen upon peculiarly profitable adaptations. The competition was among smaller groups and proceeded in a more leisurely fashion. In brief, the same consequences that typically befall the endemic organisms of isolated islands also occur to the creatures of isolated continents. Though not to the same extent as on islands, the creatures of the southern hemisphere still faced significantly fewer kinds of competition and predation, so the selective pressures were far milder.

Yes, it is certainly true that we find repeated examples of certain evolutionary parallels occurring between the northern and southern mammals. In Australia, marsupial mice, marsupial moles, rabbit-eared bandicoots, Tasmanian wolves, and marsupial lions—all developed the basic structural form of the placental counterparts. But the marsupial lion and Tasmanian wolf had fewer new places to conquer, fewer new prey to feast upon, and fewer competitors. The evolutionary environment was far more forgiving. The northern cats, by contrast, had a massive world before them, filled with all sorts of critters and climates, all especially adapted to their environments. So today we have lions, tigers, cougars, cheetahs, leopards, jaguars, bobcats, ocelots, lynxes, and margays. And when any of these felines, adapting to their new environments and trying to feed on faster and cagier prey, developed other successful traits, they then could start to dominate. The cougar prowls almost every region of the western Americas from northern Canada to the southern part of South America. The leopard ranges from Africa, throughout the Middle East, much of Asia, to the icy cold of far eastern Russia. The lion was once the most widespread of all non-human mammals—reaching the Americas, Eurasia, and Africa. We also find similar distributional empires among the canines. The jackals, coyotes, and foxes certainly do well, but the gray wolf has become particularly prolific, occurring almost everywhere in North America and Eurasia, and reaching as far south as India. Indeed, almost any mammalian species that developed some evolutionary advance in any particular region of these three continents eventually became widespread, leading to differentiation into many successful descendants. As noted, the first Carnivora ancestor produced a thriving legacy of stalkers and hunters, but even their success seems relatively mild when compared to the triumphs of the rodents (2000 species) or the bats (1000 species). (In comparison, all other mammals, put together, have divided into 1800 species.) And this is why, on the mammalian level, it would not be inaccurate to say we live in a world of rats, cats, and bats (as well as moles, shrews, monkeys, and pigs).

We can see the difference in the success rate between North and South by considering an extraordinary biogeographical event— what today is known as the "Great American Interchange." From about 35 million years ago to 3 million years ago, that is, throughout much of the ascendancy of the mammals, South America was an isolated continent—and from about 65 million years ago, it was a continent that was hard to reach. Thus, it developed a rather peculiar endemic mammalian fauna from relatively few invasions from three geographical sources. First, as noted in the previous chapter, the South American Xenarthra—sloths, anteaters, and armadillos—are sister to the Afrotheria, both groups having derived from an ancestral population divided by continental breakup, perhaps 100 million years ago. Ungulates, the ancestors to a now-extinct and diverse population of hoofed animals, entered South America about 60 million years ago, likely from North America. These South American animals, though they have now left no descendants, took many forms, superficially resembling llamas, hippos, and boars. About the same time, or not long after, marsupials with Australian ties appear in South America (presumably due to the geological connection through Antarctica). And about 30 million years ago or earlier, South America was invaded by two African mammals, monkeys and caviomorph rodents, both of which rafted across a much younger and narrower Atlantic Ocean—more of a seaway at that time than an ocean. At this point, this trickle of mammals into South America stopped for tens of millions of years, and the resulting isolation produced many peculiar South American forms—as isolation always does. The radiation among the Xenarthra produced the lumbering, elephant-sized ground sloths and the spectacular and seemingly Mesozoic glyptodonts—giant armored armadillos, as big as a small car, with spikes along their tails. The caviomorph rodents also flourished, leading to the capybara, porcupines, chinchilla, guinea pigs, and the now-extinct "Ratzillas"—*Phoberomys pattersoni*, a bison-sized, guinea pig-like creature, and its close relative, *Josephoartigasia monesi*, which weighed a metric ton and is the current record-holder for largest rodent ever. Some sizable predators evolved as well—the sabre-toothed panther-like marsupial *Thylacosmilus* ("pouch sabre") and the giant flightless terror birds, the phorusrhacids, some of which stood as large as 10 feet (3 m) tall. These birds did not become extinct until just a few million years ago and had they lived until the present day, it is likely the notion that birds evolved from dinosaurs would have occurred to biologists much earlier. The surprising size of a number of the South American plant-eaters was probably an effective enough adaptation against the relatively small number of predators there, so they remained slow and untroubled.

Thus, up until a few million years ago, South American flora and fauna were quite as distinctive as the Australia and New Zealand biota seem today—and just as vulnerable. But, in one of those colossal transformational episodes that so often grips this world, the Isthmus of Panama emerged from the sea, and South America suddenly became acquainted with the cunning, speed, teeth, and claws of the Laurasian biotic realm. As noted, biogeographers refer to the event as an "interchange," as plants and animal moved in both directions between the continents. But it is certainly clear that North America had the upper hand. Perhaps, the most obvious result we find in the fossil record is that North American meat-eaters flooded southward into this new-found rich and fleshy landscape and discovered a wealth of large and lumbering herbivores to rend and claw. As the biologist Stephen Wroe described it, "At that point, suddenly, wham, the carnivore diversity in South America goes absolutely stratospheric...You go from having a handful of not particularly big mammalian carnivores to having arguably the most extraordinary range of big carnivores in the world."[3] These predators include massive sabre-toothed cats, bears, at least six species of canines, and a 400-kg lion that was four times larger than the marsupial lion. The gargantuan sizes attained by relatives of the South American guinea pigs, armadillos, and sloths would not have been enough to protect them from the northern marauders. And the rise of the Panamanian link may be properly described as a South American disaster, not as devastating as the one that befell their Antarctic relatives, but certainly far gorier. It produced what perhaps was the bloodiest spectacle since the fall of the tyrannosaurs.

Northern herbivores, like the now-extinct American camels, moved south too, which is why we find llamas, alpacas, and other camel-like relatives in South America today. Camels also roamed across the Bering Bridge into Mongolia, Arabia, and Africa— creating an apparent disjunction of the South American and central Asian forms that would have seemed quite mysterious had we not discovered fossils of their ancestors in North America. The distributional history of tapirs is similar, moving into South America and over the Bering Bridge into Southeast Asia. They too have since gone extinct in North America.

A number of South American taxa also moved northward during this great biological swap, opossums and porcupines, for example. The giant sloth, the glyptodont, and the terror bird also made it into North America as far as Florida, but all have since gone extinct.

In general, the biological current was strongest toward the south. Roughly half of existing South American species are derived from North American forms—especially the cats, rabbits, bears, raccoons, deer, foxes, squirrels, and weasels. In contrast, only 10% of North American mammals are descended from South American ancestors. Because of the interchange, all of the ancient South American hoofed animals were replaced with northern ungulates (the deer, tapirs, and camel relatives) and nearly all species of South America's old diverse marsupial fauna have now disappeared—leaving only three forms, the wide-ranging opossums, the rat-like "shrew opossums," and the diminutive, tree-climbing "monito del monte," which looks like a mouse.

THE BLOODY FALL OF SOUTH AMERICA

Focusing more carefully on the ancient South American survivors, we find an even sharper biogeographical pattern. By and large, the pre-invasion southern mammals that were able to outlast the onslaught were those that had remained in the trees. And one place in particular in South America that provided the safest sanctuary for its unique arborealists was its rainforest. Somehow, South America's large stretch of jungle—a highly specialized, dark, three-dimensional, tree-stuffed environment—managed to counter the advantage of the massive Laurasian evolutionary machine. And in order to ascertain precisely why that is, we have to determine the origin of the most conspicuous trait of rainforests—their astonishing biological diversity.

No place on the planet contains a greater variety of species than the remarkably dense patches of water-drenched foliage known as jungles. In a single hectare of rainforest, there are likely to be 100 different kinds of giant trees and 40,000 species of invertebrates—spiders, millipedes, insects, etc. In Panama, researchers were able to collect 950 species of beetles from a single tree alone.[4] And the Amazon rainforest is home to more than 1600 species of birds. Much of the planet's biodiversity is concentrated in the 7% of the globe covered by rainforests where roughly one half of all plant and animal species reside. Why?

Biologists have argued over the cause of the extreme amounts of speciation in rainforests for decades, as well as over the reason why diversity tends to increase as one moves away from the poles and toward the tropics. In general, the closer to the equator you are, the greater the number of species surround you. This is partly because the frozen polar regions are a much more recent development, having formed in the past 30 million years. Tropical environments, by contrast, have provided a relatively stable environment for more than 100 million years (and the Amazonian rainforest itself has been around for at least the last 55 million years), giving plants and animals more time to diversify. The tropics are also a much more energy-rich region, with the warmth and moisture providing fertile regions for many kinds of plants, which in turn, attract many kinds of insects, birds, and animals.

Yet the rainforest shelters such an extraordinary density and variety of life it seems like something else is at work other than just time and fecundity. Another one of the reasons may be that rainforests are an unusual environment that is consistently shifting and splintering due to its biological nature. During dryer times, when glaciers have advanced from the poles and locked up a significant fraction of nourishing rainwater, the jungles will start to recede and fragment. This is equivalent to the formation of natural geological barriers. Many plants and trees that only thrive in extremely wet climates will not be able to colonize the drier gap—and the insects, birds, and animals that have specifically adapted to these plants and trees will also remain confined to the various patches of jungle. The resulting isolated sections would be biogeographically analogous to islands situated in proximity, like the archipelagoes described in Chapter 3. Such systems fuel biodiversity as the separated populations in each patch of rainforest begins to differentiate. And as with islands, repeated episodes of isolation-induced speciation will continue to occur as the distances between the habitats frequently change.

But perhaps the most important cause is that what the rainforest seems to lack in changing geological barriers, it more than compensates for with disruptive biological ones. To our untutored eye, a jungle may appear to be a uniform region of intertwining green, but, in reality, it is a massively dense and complicated stretch of different micro-environments. The warmth and extreme amounts of rainfall are so amenable to plant life that rainforests have become stuffed through and through with trees, vines, and shrubs of all kinds, which themselves provide a rich variety of miniature ecosystems and an often insurmountable barrier to long-distance trafficking of many other plants, mammals, reptiles, insects, spiders, birds, and even humans. The impenetrable vegetation of the rainforest is, in fact, the main reason why native New Guineans and Amazonians fractured into so many different societies—and why many of these tribes remained undiscovered by Eurasians until the twentieth century. Indeed, the rainforest in New Guinea is so incredibly dense that a massive community of native New Guineans, more than 50,000 in population all occupying the Grand Valley, remained undiscovered by Europeans until the 1938 Third Archbold Expedition (named after the leader, Richard Archbold). To this day, this remains one of the most astonishing anthropological finds in history. Though merely 115 miles from the north and south coasts of New Guinea, this prolific and seemingly stone-age society lived in complete obscurity despite the fact that Europeans had reached New Guinea in 1526 and European colonial governments had been established by 1884. Even today, smaller tribes of New Guineans (and Amazonians) are still being discovered. How could this be? As Jared Diamond explained in *The Third Chimpanzee*, the answer is "obvious as soon as one sets foot in New Guinea and tries to walk away from an established trail."[5] In 1983, in an effort to reach the Kumawa Mountains, it took Diamond and a team of twelve New Guineans two weeks to penetrate 7 miles inland. But this, he noted, was nowhere near as daunting as the attempted journey of the early-twentieth-century British Ornithologists' Union Jubilee Expedition. In January of 1910, they started their New Guinea inland trek toward mountains that were 100 miles inland. In February of the next year, they turned back, having only managed to penetrate 45 miles in thirteen months. Europeans were able to reach both the North and South Poles before they could reach the center of New Guinea.

These intimidating organic barriers coupled with the plethora of mountain ridges have not only isolated the tribes of New Guinea from the outside world, they have also separated them from each other. Each valley in New Guinea has its own clan— each with its own language, manner of dress, child-rearing tendencies, superstitions, sexual mores, diseases, and genetic abnormalities. This diversification is so great that New Guineans have developed about 1000 different languages—in contrast to the fifty that occur throughout all of Europe. Often, people born into these isolated tribes never venture further than 10 miles from their birthplace, and as one New Guinea highlander said of his life prior to their tribe's discovery in 1930: "We had not seen far places. We knew only this side of the mountains. And we thought that we were the only living people."[6]

It is not unreasonable to surmise that the same cause of the great diversity of tribes and languages in the rainforests of New Guinea—the extreme hindrance to movement caused by ecological congestion—may also have contributed to the great diversity in plants and animals in rainforests worldwide. In open continental regions, over grasslands or dry stretches, birds or insects may travel many miles before they light upon a new likely residence. But beneath the rainforest canopy, dispersal potential becomes drastically severed, and no bird, animal or insect could move very far before encountering a wealth of edible opportunities and available habitats. Jungle populations that become separated from each other by just hundreds or even tens of miles may be as effectively and permanently isolated from each other as if they were on opposite sides of the globe. Indeed, this is one of the evolutionary reasons for the very bright colors and loud calls of many rainforest plants and animals. Consider the vocalizations and vibrancy of parrots and toucans—or consider the howler monkey, the world's loudest animal. These noisy or ostentatious displays help the animals discover the whereabouts of other members of the same species within the riotous profusion of green.

Thus, for many jungle organisms, speciation can begin to occur at very short distances, and like a wildfire, this great diversity would continue to spread and fuel itself, for the different plants and animals would provide great variations in the selective pressures within all the neighboring micro-territories. These extreme isolating effects of course would only occur below or at the level of the canopy, which is that upper, oceanic layer of leaves, situated some 150 to 165 feet (45–50 m) above the ground, that forms the unbroken living roof of the rainforest.

But there is another more sparse stratum, the emergent layer, made up of the occasional behemoth tree that grows 200 to 230 feet (60–70 m) high, like the giant kapok (or silk cotton), which often breaks through the darkness of the canopy and extends upward into the open sunshine. In South America, these more sparsely situated leafy spires have become the nesting sites of the harpy eagle, the largest eagle in the world with a wing span reaching 7 feet (2 m). The harpy eagle uses its vantage point to prey upon the monkeys and sloths rustling through the meadow of canopy leaves below—a hunting dynamic that is similar to that of North American raptors perched in trees looking down for animals moving about the fields and grasslands. In the rainforest, though, this system begins at 165 feet (50 m) off the ground.

Unlike the more than 40 species of toucans that inhabit the Amazon rainforest, the range of the harpy eagle extends from Central America and all across the South American jungles; yet it has not diversified. Clearly, the freedom of movement offered above the canopy has allowed a continuous flow of genes and so the eagles remained above the rampant speciation occurring below. The giant kapok, in which these eagles so often build their nest, are similarly unconfined by their neighboring trees and have retained their powers of dispersal. Their seeds, surrounded by fluffy cotton fibers float through the air for great distances above the canopy. Like the harpy eagle, the range of the kapok extends across the rainforest with remarkably little genetic differentiation. Indeed, even the Andes have not been able to block the spread of the kapok, and within the past 15 million years, their floating puffs of seeds have even managed to cross the Atlantic, between South America and Africa.[7]

At the level of the canopy and below, a rainforest is like a massive continent's worth of ecosystems all mashed together, one overlapping another, and stacked layer upon layer from the ground floor to its leafy roof. But above the canopy, we find free open spaces where dispersal ability is no longer hindered. This has resulted in some of the more genetically homogenous creatures of the emergent layer, which remain somewhat aloof from the organic cacophony of the rainforest. This also helps explain the reason for the differences in survival rates between South America's grounded and arboreal taxa. By contrast to the rest of isolated South America, which had comparatively too few plains and open forests to provide any organic competition for the Laurasian forms, the extraordinary breadth of the neotropical jungle habitat provided an evolutionary factory that churned out arboreal mammals that could indeed survive and even out-compete the best that Laurasia had to offer. The rare characteristic of the grasping prehensile tail, so extraordinarily helpful for taxa that lived in trees, evolved no fewer than five different times among the rapidly diversifying arboreal mammals of South America—in monkeys, opossums, anteaters, climbing porcupines, and the tiny marsupial, the monito del monte. After the rise of the Panamanian Isthmus, the raccoon-like kinkajou, a northern invader, also developed a grasping tail in the Amazonian rainforest. By contrast, no descendant of North American animals that remained in the north evolved this same supportive adaptation.

Many of the successful South American climbers like the howler monkeys, the spider monkeys, marmosets, two-toed sloths, woolly opossums, and woolly mouse-opossums could not travel deep into North America because they are mostly restricted to the jungles, but, then again, the new influx of northern carnivores could not readily get at them either. The cats did indeed have certain advantages that helped them succeed in the tropical forest—leading to the jaguars, ocelots, and margays—but they could not run through their prey as did their sister cats working the grasslands. They had to adapt to the peculiar ways of the jungle, slowly adjusting to arboreal prey that had an evolutionary head start. The rainforest jaguar, which is much smaller than those found in open terrain, can only reach monkeys when they come down to the lower branches. Since the feline invasion, only the tiny margay, a sister species of the ocelot, has managed to become a true tree-dwelling specialist, living with and preying on the porcupines and opossums in the canopy. They are the smallest of the wild cats, not much larger than a domesticated house cat. They have very large paws, and their ankles rotate 180°, allowing them to climb down trees head-first, like a squirrel.

The South American arboreal fauna not only had a head start on the predators, they were also far more capable of withstanding the northern competition of similar creatures than were their ground-based counterparts. The descendants of the northern raccoons and squirrels, which remain the relatively lonely and undisputed furry masters of North American treetops, have become just two more mammals in the canopy. They have not been able to dominate south of Panama. By contrast, the top ground-based predators of South America lost out to the Carnivora, and the North American hoofed animals entirely replaced all the archaic southern ungulates. From an ecological viewpoint, the fall of South America may be considered a tragedy—a massive introduction of countless invasive species to an isolated and fragile ecosystem, resulting in a bloody rout of its endemic terrestrial fauna. But at least we may take heart that the wild and noisy layer of the rainforest canopy is still playing ancient South American music.

We now should be able to see the vague outlines of the terrestrial biogeography of the vertebrate world, especially involving the mammals. On the remote oceanic islands, we have the great dispersers—those plants and birds that can travel by wind, waves, or water. With very few or perhaps even no exceptions, nothing native on these lonely isles has four legs. Nearer the continents, the more daring vertebrate travelers begin to appear on the volcanic islands, the rafting lizards and floating tortoises.

THE BLOODY FALL OF SOUTH AMERICA

On the continents, the major event that has transformed the entire biotic organization of the world was the formation of the circum-Antarctic rift system, which generated the northward drift of the Gondwanan landmasses and stranded them in the ocean-dominated hemisphere. The result has been extreme endemism in the south—the peculiarizing of so much of their flora and fauna. The isolation has also helped provide sanctuary for plants and animals dating from the time of the Gondwanan radiation. Thus, it is in the south that we are more likely to find the ancients and the exotics—those endemics that seem so delightfully strange to people of the northern hemisphere. When Americans visit England or France, they are not typically inspired by the aboriginal plants and animals, which are very similar to eastern American taxa. But when they travel to Madagascar, Australia, or New Zealand, they feel like they have entered an alien biological realm.

The continental connection between North America and Eurasia—and their long association with Africa—increased the chances that any successful mammal evolving anywhere among these continents would have a chance to spread, diversify, and produce more beneficial adaptations. This resulted in fiercer evolutionary struggles, as more skillful mammals fought for resources, and more effective carnivores sought to capture fleeter and warier prey. The result has been that the large-vertebrate world has become dominated by the northern placentals. Even northern groups that became isolated in a southern continent relatively early in history could not compete against the descendants of ancestors that remained in Laurasia. The one exception to this rule appears to have occurred in the tropical rainforest, those self-perpetuating engines of biodiversity, which, far more than any other isolated region, have produced animals that exhibited greater resistance to extinction in the face of northern invaders. The evolutionary engine that was in idle through other parts of the southern hemisphere remained steaming and straining in the rainforest—and the intensive selective pressures were maintained. But excepting these jungle holdouts, this has become a northern mammalian world.

And all of this—the icy death of Antarctica, the preservation of ancients in Australasia, the domination of the northern mammals, the fall of South America, indeed, this entire global organic pattern—can be traced to a series of volcanic cracks in the ocean floor that started to surround Antarctica more than 80 million years ago.

CHAPTER 6

Enchanted Waters

Biogeography among the life aquatic

A t first, it almost seems like there could be no real biogeography of the seas. Many of us have gotten our glimpses of ocean life through television nature specials, the occasional trip to the local aquarium, or whale-watching cruises, and it would seem that the watery majority of our planet is just a great big cluster-jumble of fish, squid, crustaceans, jellyfish, bivalves, and sea mammals swimming, floating, and crawling this way and that, completely unobstructed by the same type of barriers that impede the land-lovers. Tuna, sharks, swordfish, and dolphins swim all over the Atlantic, Indian, and Pacific Oceans. We find killer whales stalking seals in the Arctic and killer whales hunting penguins in the Antarctic and killer whales practically everywhere in between. After all, what could possibly stop these creatures? Continental plants and animals often find themselves corralled by mountains and lakes, deserts and jungles. And their populations are routinely divided by the floods, riftings, and upthrusts that transform our landscapes. Yet, all this terrestrial turmoil that is such a significant driver of biodiversity on land is entirely irrelevant to the fish and mammals of the ocean.

So it would seem that this would to have to be a very brief chapter.

But as we shall see, the same biogeographical principles that apply to animals above the waves also apply to those below. Certainly, many marine species that move so easily through the seemingly featureless depths do not, at first, appear to submit to the curiosity of mappers and plotters. But, after careful scrutiny, we find that even with the greatest of swimmers, organic patterns do emerge and that those patterns coincide with geological or oceanic processes.

As noted in the last chapter, the "Great American Interchange" refers to the biological swap that occurred between North and South America when the Panamanian Isthmus finally pushed above the sea and connected the New World continents. But from the perspective of the marine denizens of the Caribbean and East Pacific, this exact same event would be considered the "Great American Divide"—as, at that point, sea life in these marine regions suddenly became separated. For residents of the oceanic realm, the formation of a continental barrier like the Isthmus of Panama is analogous to the continental breakup discussed in Chapters 4 and 5. Just as the opening of the Atlantic necessarily led to wide-scale differentiation into African and South American varieties of numerous organic groups, the same thing occurred with the marine flora and fauna now divided by the Panamanian bridge. What were once unified populations of fish, crustaceans, clams, etc., have now diversified into East Pacific and Caribbean-Atlantic species.

A recent analysis by a research team headed by Ada Kaliszewska from Harvard's evolutionary biology department confirmed that right whales, like countless other fish and bivalves, had indeed become split by the Panamanian divide five to six million years ago—leading them to diverge into North Pacific and North Atlantic species. The rising isthmus also changed ocean currents causing another population of whales to become isolated by another barrier—one that is inconspicuous to us but that is very dissuasive to right whales: the warm equatorial waters. Right whales have lots of blubber to protect them from colder waters, but they find the bathwater temperatures around the equator intolerable. Thus, we have three species of right whales that have diversified into their present forms due to their fragmentation by barriers (Figure 9). As above the sea, so below.

One of the more fascinating elements of Kaliszewska's analysis was that much of this information was gleaned not by capturing and studying the whales themselves—but simply the whale lice that pester them. "Whale lice," despite their name, are not insects but crustaceans, half an inch in length, that look a little like tiny crabs. Colonies of these light-colored crustaceans feed on the whales' "callosities"—which are the raised fleshy patches that cover their heads. As with fingerprints on humans, these raised sections of skin form a design unique to each individual, and the fidelity of these whale lice to these patches makes the callosity-patterns bright and conspicuous, helping scientists to identify individuals.

Like real lice, these cetacean parasites cling to their host and eat dead skin. Also, like lice they confine themselves to the same species—just as bird lice and human lice do. The whale lice, though they can crawl about on their hosts, cannot swim. So they can only move from whale to whale when they are in contact, typically during mating.

Kaliszewska and her colleagues were able to use molecular analyses of the various populations of lice to determine patterns

9. The distribution of the three species of right whales.

of relationships and timing of divergence among these whale riders, which, they knew, would correspond to the speciation of their giant hosts.[1] They discovered that the main differentiation of the Atlantic and Pacific groups took place during the rise of the Panamanian Isthmus. They also observed that the southern hemisphere whale lice became separated from their northern Pacific sisters a little after their isolation from the north Atlantic forms, indicating separation by the more recently developed warm-current barrier of the tropics. Finally, some small fraction of the whale lice in the northern Pacific also appears to be very closely related to southern Pacific relatives, far closer than most of the other northern Pacific whale lice, suggesting that, after the three groups of whales had divided, at least one confused or desperate whale managed to cross the equatorial boundary into the north Pacific at some time when the currents were coolest—perhaps one to two million years ago.

Still, the extent of right whale distribution is quite large—and few continental organisms cover such a broad area without having diversified into many more forms than three (humans, as we shall see in Chapter 7, present a notable exception). This confirms the rather obvious fact that, from the point of view of the greatest swimmers, the ocean environment is considerably more uniform than the continents. This is why marine mammals, despite their extraordinary success, only constitute 2.5% of mammalian species. Although they populate all regions of the global oceans, they have not diversified to the same extent as have their continental counterparts.

Tuna, swordfish, sailfish, and marlins are other great conquerors of the oceans—a fact aided by a series of streamlined adaptations that have graced them with their celebrated ease and swiftness in the sea. Tuna have fins that they fold into a groove along their back so as to provide no additional resistance when they have the need for speed. The lenses of the eyes of swordfish do not bulge outward but run flush with the smooth head, and, as with the tuna, their tail tapers to a crescent. But it is perhaps the swordfish's close relative, the sailfish, that has achieved the hydrodynamic ideal. It is the speediest thing in the sea, having been measured to swim at 68 miles per hour ($\sim$109 km/h)—about as fast as the cheetah is on land.

All of these fish migrate through the open and unobstructed ocean with little effort. But even among these fish, we have found some surprising differentiations—even shared genetic patterns. Specifically, the big eye tuna, swordfish, and blue marlin of the Atlantic Ocean all comprise two different clades that are currently mixed together. These groups are not precisely different species, as they still interbreed, but, within each fish, there coexists two Atlantic varieties that are genetically distinguishable. In each case, one of the clades is exclusive to the Atlantic, while the other has a more cosmopolitan distribution, appearing in the Atlantic, Indian, and Pacific Oceans. Such a pattern suggests some sort of "vicariant" explanation—a single, common geological or oceanographic cause that resulted in this shared genetic structure.

The Agulhas current provides that explanation. This strong oceanic corridor flows from the Indian Ocean around Southern Africa and into the Atlantic, making it much easier for the big eye tuna, swordfish, and blue marlin of the Indian Ocean to reach the Atlantic rather than vice versa.[2] Evidently, this current is dependent on glacial cycles, becoming much weaker when the Earth is colder and more ice is locked up along the poles. When the Agulhas current is weak, the Atlantic and Indo-Pacific forms become temporarily isolated, leading to their genetic differentiation. When the Agulhas returns, the Indian Ocean fish are able to flow with it into the Atlantic.

Ironically, the most persistent ocean migrator appears to be the slowest (and indeed the largest) fish in the ocean. The massive whale shark, reaching 60 feet (18 m) in length and a weight of 40 tons, moves leisurely through the seas feeding on the worldwide nutritious floating stew of tiny crustaceans, worms, mollusks, and fish eggs known as "plankton." Researchers have recently conducted a genetic analysis on whale sharks from different regions around the world and discovered that despite their unhurried lifestyle, they are truly a global species. Whether you find them in the Gulf of Mexico or near West Africa or off the coast of New Zealand, they are the same species with very little genetic differentiation. The whale shark is one of those few creatures that simply knows no barriers.

Such genetic homogeneity applies mainly to the long-distance travelers, the open-ocean swimmers, while much of the spectacular variety of ocean life remains confined to the coasts or along the bottom. The intertidal and coastal creatures are just like their continental counterparts. Some of them can disperse long distances (usually because they have a freely floating larval stage that allows them to be swept along in various currents) while others are very sedentary. The less mobile sea creatures are stuck to the continents, just as the land-dwellers are, and their genetic relationships with geographically distant species betray the past locations of the land that supports them.

Sea catfishes, for example, are confined to continental shelves both as adults and in their early stages. Not long after fertilization, the males carry the eggs in their mouths for months until they are ready to hatch. Thus, we find in one clade of these catfish the familiar South African–South American split caused by the opening of the Atlantic—but with one interesting difference. The South American form lives along the *west* coast of South America, near Peru. Why? Figure 10, with the black-line representing the distribution of the group, suggests the likely history.

This particular family of sea catfishes lived along the southern ends of both continents when they were juxtaposed. But after the Atlantic split, the currents near the southern end of South America became too cold, pushing the catfish toward the warmer waters up the western coast. It is also possible that the range of the fish had originally extended up the western coast of South America, and they subsequently died out in the southern regions with the cooling of their waters.

The differences in the type of adaptations required for marine animals to survive at different depths add another layer of complexity to marine biogeography. While for the distributions of land creatures we think mostly in two dimensions, north–south or east–west, the oceans add a third.

The top 660-foot (200-m) layer of the ocean is known as the euphotic (*good light*) or sometimes just photic zone. Here is the lush, productive realm of the ocean where everything directly dependent on photosynthesis to make energy—the plants, phytoplankton, and cyanobacteria—must remain. This energy feeds many groups of herbivorous fish, who, in turn, feed the piscivores. Only 1% of the sunlight that reaches the surface can penetrate beyond the euphotic zone, and everything below it must live in a perpetual twilight. Fish have adapted to the general darkness of much of the ocean by developing what is known as a lateral line system—lines of sensory hair-cells running from head to tail that detect differences in pressure. They are analogous in

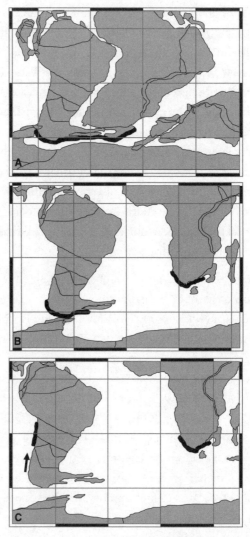

10. Biogeographical history of a sea catfish group, separated by the opening of the Atlantic. See text for more detailed description.

structure to the ear, which, in a similar fashion, uses hair-cells to perceive the repeated cycles of pressure variations known as sound waves. This similarity, of course, implies an evolutionary connection.

Fish are able to use the lateral line system to sense objects moving about them in any direction. Also, when fish swim they push water before them in a type of bow wave that reflects off any other solid object. These watery echoes then rush over their hair-cells, alerting them to the oncoming object. This is why fish do not bump into the glass walls of the aquarium. This is also why fish are able to adapt to the lightless conditions of caves or the ocean deep. Blind species of cave fish or bottom dwellers are not helplessly moving through a shapeless world, but have a very specific sense of their surroundings. Fish are also able to perceive the bow waves of approaching predators with their specialized rows of cells. If you have seen schools of fish all change directions at the same time, you may have gotten the feeling that they are all moving as a single unit and communicating with each other through some sort of extrasensory perception. And in a way, this is sort of true, for the lateral line system that allows this seemingly immediate communication of movement through the water is a sense that is beyond our five—though still a purely natural, materialistic process.

Dolphins and whales, having evolved from mammals, lack this important undersea sense. But many of them have developed a substitute that is very similar—echolocation. These sea mammals continuously emit sound waves, typically by clicking, and then use the echoes reflecting off objects to develop a three-dimensional image of the world. Many nocturnal and cave-dwelling bats employ the same system. From our point of view, echolocation is somewhat like using a flashlight in the dark. The emitted waves of the flashlight reflect off various objects and then press against the back of our eyes. Our brains automatically transform the wave patterns into a three-dimensional image of the surrounding world. Similarly, bats and dolphins emit clicks in order to "light up" the dark world around them—and their brains automatically transform the reflected waves into an image.

Seals lack both the lateral line system and echolocation, so they have developed something else to help them catch fish in the dark. They instead have whiskers that are very sensitive to the hydrodynamic trails left by their prey. All of this helps give force to the prescient insight of the classical Greek philosopher Democritus that all senses are forms of touch. That is, you must be physically struck with something material in order to be able to sense distant objects—whether it is molecules of a rose entering a nose or the reflected wavelengths from its petals, which we see as red, or the sound waves of the bee, which we hear as a buzz. We are all like fish in an ocean, being constantly bathed in the material evidence of the surrounding presences—all of us having evolved from those original rudimentary life-forms that first began to get a vague sense of the dark world about them.

Among sea life, the bottom or near-bottom dwellers may be divided into five basic categories. The shallowest types are the intertidal creatures, the taxa that you often see at the beach that live between the high and low tides like crabs, sea stars, snails, sea anemones, and mussels. Then, in descending order of depth, there are the organisms that live along the continental shelves, the deeper taxa that we find along the continental slope toward the abyss, the alien abyssal forms on the deep ocean bottom, and then the extremely simple life of the hadal zone that are able to survive the crushing, cold, blackness of the deepest ocean trenches. As various types of plants and animals can only inhabit certain elevations along mountains, various types of bottom dwellers can only inhabit certain depths.

At the deepest point measured, the Challenger Deep, 7 miles (11 km) below the surface, the pressure is so intense that no hard-shelled creature can possibly survive. Here we find only some forms of foraminifera, tiny soft-shelled, single-celled, amoeba-like globs of living material that are also abundant in the upper sunlit layer of the ocean. The foraminifera are not quite plants or animals or fungi but part of a separate kingdom—the protists, which also includes algae and slime molds. They have been around in similar form since the Cambrian. The foraminifera at the lightless bottom of the Challenger Deep are most closely related to the foraminifera that live at a less extreme depth— say, 3 miles (5 km) below the surface. In the deep sea trenches, the distributional patterns run vertically, going from deep to deeper.

The mid-ocean ridges, those same volcanic canyons where new seafloor is produced, have also become a place of biological fascination. In 1977, scientists in the famous deep-diving submarine *The Alvin* discovered to their astonishment that the hydrothermal vents that often surround these abyssal regions sustained a diverse and previously unknown group of sea creatures. Hydrothermal vents are like deep sea geysers that spew hot mineral-laden seawater into the oceans. A certain type of bacteria feast and thrive in these sulfur-rich environments, converting the emanating chemicals into their living tissue. They, in turn, have become the solid base of a food chain—the microscopic foundation of a complicated, self-sustaining ecosystem that derives its energy exclusively from chemicals (chemosynthesis) rather than light (photosynthesis). The larger, mysterious creatures of this alien realm include tubeworms that can be 10 to 13 feet (3–4 m) long, eyeless shrimp, giant clams the size of footballs, and blind white crabs. The worms are particularly novel. They have no mouths or guts, and absorb the bacteria through feathery appendages. The bacteria live inside them and secrete the essential nutrients the worms need to live. This is an example of an "obligate" relationship. Without the bacteria, the worms cannot survive. One particularly strange species of worm, the Pompeii worm, earned its volcanic name because it lives on the side of the vent itself, rather than nearby, making it the most heat-tolerant complex animal known to science, surviving temperatures up to 176 °F (or 80 °C).

The peculiarly self-contained nature of these vent systems, isolated as they have become from the dominant light-based world-ecosystem, raises some biogeographical issues that have recently been addressed by the vent researchers, C. Mary Fowler and Verena Tunnicliffe.[3] An analysis of the vent fauna from different sections of the global ridge system confirms that the biological similarity between the fauna of different vent systems does not correspond with the shortest-ocean route (as the whale-shark swims, so to speak) but corresponds to the shortest distance along the ridges (Figure 11).

In other words, the vent systems that run along the rifts provide a biological corridor and the larvae of these creatures carry the genes along it, with few or no cross-ocean short-cuts from one part of the ridge to another. Once again, vicariance, the isolation of taxa by the formation of geological barriers, plays an important role. About 30 million years ago, the northeast Pacific spreading ridge just off the coast of California

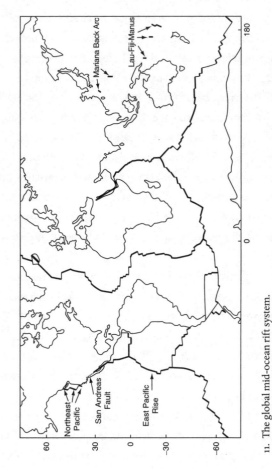

11. The global mid-ocean rift system.
Hydrothermal vent systems tend to follow the chain of seafloor spreading ridges that run throughout the world's oceans.

became thrust beneath the North American plate by a tectonic process geophysicists call "subduction." Currently, the San Andreas Fault cuts through the western part of California from Cape Mendocino near San Francisco to the Gulf of California, where it once again runs into a seafloor spreading ridge of the global rift system. Everything west of this fault line, including Los Angeles, San Diego, and the Baja Peninsula is no longer part of the North American Plate but has now become part of the Pacific Plate. The friction and strain caused by the two plates sliding past each other generates the California earthquakes along the San Andreas Fault, which represents a sort of continental manifestation of this swallowed east Pacific plate divide. More relevant to the creatures of the regional vent systems, the subduction of this east Pacific ridge separated the vent communities of the northeast Pacific from those of the east Pacific ridge (and the rest of the world), leading to the differentiation of these vent communities.

But there's a distributional mystery. What about the isolated ridges in the west Pacific? Where did the fauna of these far-flung regions originate? According to the analysis of Fowler and Tunnicliffe, many of these lonely creatures are most closely related to the isolated taxa of the northeast Pacific ridge, which again coincides with the history of the Pacific. The spreading ridge that produced the north Pacific actually ran the entire width of the basin just south of the Bering–Alaskan–Aleutian region, connecting the hydrothermal societies of the Mariana basin with those in the northeast Pacific. This ridge has also been subducted, isolating the taxa on opposite sides of the Pacific.

Biogeographers, familiar with the subject as they are, tend to discuss such deadly and transformational geological events like the subduction of vent communities with a clinical detachment.

But in order to have a realistic assessment of life history, it is important to recognize the catastrophes continuously inflicted upon the organic layer that stretches over this restless planet. As with the freezing of Antarctica or the rise of the Isthmus of Panama, the slow descent of these vibrant communities into the deep and crushing abyss presents a solemn vision. Many other creatures of this Earth also inhabit fated regions—like the flora and fauna of the always-sinking oceanic islands or those parasites that, like whale lice, must die with their host. In these instances, long-term survival of descendents depends exclusively on a never-ending migration. Such species must keep moving or die, keep spreading along a conveyor belt that carries them steadily toward oblivion.

The freshwater arteries and veins of the continents flow into the great oceans, providing a biogeographical connection between the fauna of rivers and lakes and those of the great seas. Unlike the oceans, however, intercontinental waterways are far more ephemeral. Lakes dry up and rivers disappear or become rerouted. As always, a study of distributions can help recover these geological histories.

Currently, the water of the Amazon River begins in the streams and tributaries of the Andes on the western side of South America; collects in the giant, greenish eastward moving river; flows through the lush and unruly Amazonian rainforest, stretching across the continent; and finally gushes into the Atlantic. This giant river is home to a great diversity of freshwater stingrays, and, for many years, it was believed that the ancestors of these stingrays must have entered the Amazon from the Atlantic. Recent analyses, however, have shown that the freshwater stingrays are really more closely related to a Pacific group. This includes one study conducted by Daniel R. Brooks, who determined that the parasitic worms that infect the fresh-water stingrays are most closely related to those that infect the Pacific stingrays.[4] (A renowned biogeographer, Brooks has long observed that the study of parasites can help resolve many distributional mysteries, and now other researchers are following Brooks' example—as demonstrated by the whale lice study mentioned earlier.)

The stingray analysis helped uncover secrets of the history of the Amazon River that have since been confirmed. Prior to the rise of the Andes, the Amazon River actually flowed westward and emptied into the Pacific.[5] The formation of the Andes eventually blocked the westward flow, creating for a time a combination freshwater and saltwater sea between two elevated regions. As this sea continued to freshen, the stingrays continued to adapt to a freshwater lifestyle. The steady rise of the Andes eventually reversed the course of the Amazon, forcing it to flow eastward and empty into the Atlantic. This is consistent with the fact that freshwater stingray fossils have been found on the western side of South America in Peru and Columbia. Other Amazon River species that likely have a similar Pacific origin include crabs, pufferfish, needlefish, anchovies, and oysters.[6] The same tale may also describe the evolution of Amazonian river dolphins, whose closest relatives outside of South America are in China, suggesting both groups of river dolphins derived from a Pacific ancestor.

River dolphins have the distinction of being one of the few large, warm-blooded, freshwater fish-eaters in the world. Although sizeable mammalian predators dominate the land and prowl the oceans, they have been unable to conquer freshwater habitats with the same ease. Cold-blooded alligators and crocodiles float along the rivers of every continent except Antarctica. Giant snapping turtles lurk in the waters of the Americas. And many kinds of freshwater fish, like the 440-lb. (200-kg) Nile perch or the 10-foot (3-m)-long alligator gar, have reached great sizes. But where are the large flesh-eating mammals of the rivers and lakes?

Currently, no scientific consensus has been reached on the question. But Chris Lavers has suggested two likely causes for the dearth of large, freshwater, mammalian carnivores.[7] The first is that the cold-blooded nature of alligators and turtles has allowed them to survive the instability of freshwater environs. When a river dries up, they either walk to another one—or, due to their resistance to starvation, they can dig themselves into the mud and wait, sometimes for months, for the rainy return of wetlands. Lavers suggests crocs might even be able to survive a couple of years entombed like that. River dolphins and manatees (which are herbivorous), by contrast, can only live in the largest and most stable systems.

The other reason is simply that the extremely successful crocodiles have prevented it. Recall that the dramatic radiation of mammals followed the extinction of the dinosaurs and marine reptiles 65 million years ago, as they expanded into the global ecological vacuum of the land and sea. But crocs, 'gators, and turtles of the freshwater systems survived the extinction event, and so mammals could not penetrate these regions with similar success.

One smaller mammal does rule an isolated lake, the nerpa freshwater seal of Lake Baikal, and its presence is as curious as the remarkable Siberian waters that it inhabits. Baikal is the deepest and most voluminous lake in the world by far, descending in some places more than a mile (1.6 km). It contains 20% of all the world's (unfrozen) freshwater—more water than all five Great Lakes of North America combined. Because tiny crustaceans consume much of the algae and help filter its water, Baikal is renowned for its clarity, allowing one to peer as much as 130 feet (40 m) downward. This makes some tourists feel queasy when they look down over the side of the boat and peer into its depths.

This Russian jewel is also the world's oldest lake, having formed more than 20 million years ago, but it may be more accurate to think of it as the world's youngest ocean. Baikal has formed over a continental rift system that is breaking Russia apart, similar to the second most voluminous lake in the world, Lake Tanganyika, which sits upon the East African rift system. Both rift lakes have hydrothermal vents, much like the ones we find at the bottom of the oceans.

Since the lake is so isolated, situated more than 930 miles (1500 km) from the Arctic Ocean in the center of the East Asian landmass, it is essentially identical, from a biogeographical perspective, to an oceanic island. Typically, creatures can only reach the lake through its rivers and often must go against the current. Thus, it receives few new immigrants, and many of its flora and fauna are descended from inhabitants that had already lived in local waters. Of its nearly 1550 species of animals and 500 plants, 80% are found nowhere else. And endemism is particularly high (99%) among its diverse group of gammarid crustaceans, which look like tiny shrimp. Lake Baikal also contains a particularly impressive variety of colorful worms, one as long as 12 inches (30 cm).

Its star resident, though, is its mysterious seal—whose origin has now been determined by a molecular analysis of seal relationships. According to the results, the nerpa likely diverged from an Arctic ancestor about 2 to 3 million years ago—and almost certainly reached their current home at some point when the Arctic Ocean flooded the Asian continent.[8]

Thus, we discover within this landlocked lake all of the same biogeographical principles that shape the life of an oceanic island. While Lake Baikal is a hot spot of organic peculiarity, boasting more than 1600 plants and animals found nowhere else in the world, the wilderness surrounding it is not quite so exotic, occupied by bears, deer, and wolves that are quite common throughout Eurasia. Similarly, the Galápagos Islands form another oasis of diverse and specialized taxa, thrust up in the middle of a sea of biological ordinariness, that is, among the more familiar and widespread sea life of the east Pacific Ocean. For those who fancy ecological rarity, Galápagos and Baikal are like solitary diamonds among vast fields of coal.

Likewise, finding a typically marine mammal isolated in Lake Baikal is like finding a land-based mammal stranded on an oceanic island. As with the Channel Island mammoths, which reached their island homes during lower sea levels that narrowed the gap between the continent and island, the seals of Lake Baikal required a marine transgression to bring the Arctic Ocean closer to the lake. Even the small size of the nerpa seals, which must make do with fewer resources than their oceanic brethren, is analogous to the dwarfing that is common to island mammals.

It is fitting to save the biogeography of killer whales for the last section of this chapter because the subject naturally leads us to the next—the chapter on human beings. Despite their names, killer whales are actually part of the dolphin family. They are one of the cleverest of mammals—and their complicated patterns of behavior and remarkable ability to learn have produced a new sort of genetic barrier that also helps divide human populations. Currently, the obstacles to gene flow that are leading to the partitioning of orcas are neither geological nor oceanographic, but cultural. Incredibly, genetic differentiation among orcas is being maintained by something akin to societal traditions.

For many years, marine biologists observed that the killer whales that prowled the western coast of the United States and Canada, for the most part, moved in large pods of 10 to 25 individuals, typically headed by a mother and comprised all of her descendants. The behavior and migration patterns of these pods were quite predictable. They always returned to the same hunting locations every summer and always sought the same prey. Yet, occasionally, researchers would spy other smaller groups of the black-and-white mammals in the region, with 2 to 5 individuals travelling together, not strictly based on family lines, and their movements seemed irregular. The scientists speculated that these were ostracized orcas and gave them the name "transients" to distinguish them from the larger more commonly observed matriarchal "resident" populations.

Eventually, it became apparent that the two different groups of killer whales did not mingle and that they had some physical differences. The tip of the dorsal fin is rounded in residents, pointed for transients. They also differ in the shape and shading of the white patch behind the dorsal fin. But the greatest distinction between the two is diet, which in turn molds their lifestyle and social structure. The resident orcas love fish and particularly salmon, and their migration patterns are timed with that of their quarry. In mid-June, great armies of salmon from the northeast Pacific invade the Johnstone Strait just north of Vancouver Island on their breeding run into the innumerable rivers and streams of southwest Canada. And pods of resident whales arrive to the strait just prior to this entry and establish a fierce and toothy gauntlet. The killer whales are very chatty as they move from feeding spot to feeding spot, communicating the location of their prey with clicks.

Transients, by contrast, ignore fish altogether, much preferring the taste of mammalian blood—especially harbor seals, sea lions, porpoises, and most any kind of whale, including sperm, beluga, humpback, and gray. If it is in the water and has lungs, the transients want to eat it. They have even been seen attacking river otters, as well as deer and moose that have ended up in the sea at the wrong time. Their fidelity to this diet of milk-producers is remarkably strong. In 1970, three captured transient killer whales refused to eat the fish provided for them for 75 days, when one finally died. On the 79th day, the other two began to eat salmon, but switched to a marine-mammal diet when returned to the sea.[9]

This dietary dedication has led to very different foraging strategies. Unlike their piscivorous counterparts, transients mostly hunt in silence so as not to alert their sonar-sensitive targets. But once they have gored their victim, they vocalize enthusiastically. Seals, in fact, can tell the difference between the calls of resident and transient killer whales—and have become quite accustomed to the constant chatter among the salmon-eaters, carelessly continuing their marine activities whenever they swim by. But the call of the transients strikes terror in seals and sends them scurrying to shore or diving deep. Scientists also believe that killer whales have altered the migration patterns of baleen and sperm whales—and that pregnant female gray whales make their extraordinary trek from the Bering and Chukchi seas between Alaska and Siberia to warm-water lagoons of Mexico's Baja Peninsula just to keep their newborn calves safe from the jaws of transient orcas.

One of the more fascinating examples of ingenious animal behavior of which I am aware involves transient killer whales trying to get at a seal who had sought safety atop an ice floe. First, the orcas nosed the floe out to sea, then repeatedly charged the frozen platform in groups of two or three, swimming very near the surface, and creating a massive bow wave that washed over the ice and pushed the seal very near the edge. Orcas would then surface on the other side, hoping to grab their prey. Killer whales would make their runs from different sides of the floe apparently to ensure their waves would hit the seal broadside. Eventually, their strategy worked and the seal was washed over the side. But the killer whales did not devour their victim. Instead, they merely "played" with it, chasing it around, and, incredibly, when finished, they pushed the seal gently back onto the ice.[10] Evidently, it was a training run, possibly to educate the juveniles, and researchers believe this wave-hunting technique has been passed down from generation to generation.

Another recently discovered hunting technique more fully demonstrates the educational tendencies of killer whales. Long line fisheries operated in Prince William Sound for many years before a few killer whales from a resident pod learned to steal the fish from the lines. This behavior was transmitted to the rest of the group, then quickly became spread from pod to pod. Once the technique became ingrained, the fishermen found it very difficult to deter the killer whales. Even shooting them or using underwater explosives did not stop their underwater thievery. It seems that innovation among killer whales occurs rarely and slowly, but once it does, the new methods spread very quickly. As we will see in the next chapter, a similar claim can be made about the spread of ideas among human beings. As with acres of dry kindling, one spark can set the whole area ablaze. But you have to wait for that first spark.

In the early 1990s, a third group of killer whales were discovered in giant pods, some 30 to 60 individuals in a group, cruising the open ocean near the edge of the continental shelves of western North America. The "offshores," as they are called, have not been studied to the same detail as the residents and transients, but it appears they mostly live on sharks and smaller fish.

These distinctions here only refer to the northeast Pacific orcas, which are so much more accessible to American biologists. But it seems like similar patterns also occur throughout the other oceans. Three more groups of orcas roam the southern hemisphere—one that focuses on the Antarctic toothfish, a second that seems exclusively to prefer Antarctic minke whales, and a third that has developed a taste for sea lion pups. This latter group has been observed working the shores of Argentina and has become renowned for the extraordinary practice of nearly beaching themselves to bite a pup by its tail, slap it on the rocky shore a few times with a few violent shakes of its head, and then fidget itself back into the water. Like their north Pacific brethren, they too will play with the sea lion pups, seemingly refining their techniques. And, at times, if they are well fed, they will bring the last pup to shore—just as the wave-hunters saved the seal they had knocked from the flow. This would seem to be a conservational impulse. If they are no longer hungry, it is better to save these mammals for another day—rather than let them die and go to waste. Even if they never catch the seal again, it is always helpful to boost the numbers of your prey population—as fisher-men will often throw the smaller fish back into the sea—though I am not suggesting that orcas are thinking that far ahead.

In the northeast Pacific, the domain ranges of the different groups of whales are thoroughly intermixed. You will often find groups of offshores or transients moving between two resident pods. Yet a recent genetic analysis has shown that neighbor-ing populations of two different types will still not interbreed. Instead, transients, residents, and offshores, even when migrating to different pods, choose to join and mate only with members of their own type.[11] With killer whales, cultural similarity governs gene flow to a greater extent than geographical proximity.

This is akin to the genetic isolation promoted by the imprinting of the birdsong of the Galápagos finches discussed in Chapter 3. When two populations of finches are separated for a brief time, their mating songs start to differentiate. This helps erect a mating barrier between once-separated populations that is maintained even when the finches are reunited and still technically part of the same species. But there is a difference between the imprinting of bird song on finch daughters, which is purely instinctive, and the group-specific diets, vocal habits, social hierarchy, migrat-ing patterns, and foraging strategies of the killer whales, which are learned. The behavior patterns that divide killer whales are really more like human cultural barriers rather than instinctive ones, making the transients and residents somewhat analogous to ethnic groups. The genetic differences between the whales are quite small, but their "cultural" disparity is immense, involving deviations in diet, social structure, perhaps even language. This, in turn, is helping promote their genetic diversity. A similar phe-nomenon, as we shall see in the next chapter, also occurs with people.

Researchers often tend to draw a stark distinction between learned activities and innate desires, but these two behavioral guides are really intricately connected. Instincts are like computer hardware, the fixed characteristics directly installed into the computer design. And learning is like software, the programs you plug into the computer after you own it. An exclusively instinctive animal would be like a computer with no inputs. All its functioning would come from its initial configuration. A highly educable creature is like a typical computer that has been specifically built to accept software—and then runs accordingly. The relevant point is that the computer has to be built to accept software. In other words, killer whales are innately predisposed to adhere staunchly to the lessons taught by the wiser individuals of the pod. It is in the nature of the killer whales to respond so forcefully to nurture—a trait that makes them the star attractions at Sea World.

Sometimes, as is apparent by the mammal-eating killer whale that starved rather than switch to fish, their fidelity to prior teachings and experiences is so strong that it even seems to override their instinct for self-preservation. And this too is a troublesome fault in human beings. Like the killer whales, we are easily molded, but then we harden. It is in our genetic makeup to cherish, adhere to, and pass on the traditions we learn from our parents or elders. But at times even irrational and dangerous ideas can become so ingrained and irresistible that it dominates all other instincts—as when a suicide bomber takes his life and that of many others.

Here, it should be noted, I do not suggest that either human beings or killer whales, educable as they and we are, are infinitely malleable. We are not blank slates upon which anything can be written. For it is quite clear that certain lessons are much easier to learn than others. Killer whales are predisposed to learning their hunting techniques in the same way we seem predisposed to learn a fear of snakes or a distrust of foreigners. Thus, orcas, while initially confusing to the biogeographer, provide invaluable information for the sociobiologist. Their scholastic nature, their complicated social dynamics, and their culturally delineated genetic structure all give us insight into the evolutionary origins of analogous qualities in humans.

CHAPTER 7

The Battle Over Eden

The controversy over the biogeographical history of the human race

E. O. Wilson devoted almost his entire book *Sociobiology: The New Synthesis* (1975) to using evolution to explain the behavior and social dynamics of insects and non-human animals. *Almost.* In the last chapter, Wilson then made the seemingly obvious point that those same Darwinian principles that have governed the development of societies of wasps, elephants, ants, and wolves must also have exerted some influence on human social dynamics.

And that was when the controversy began.

Many politically minded academics expressed an almost Victorian indignation at Wilson's implication that it was not just our bodies that evolved—but our minds as well. And for simply underscoring the straightforward ramifications of Darwin's discovery, Wilson was denounced by some as spewing racist propaganda. At one of his lectures, demonstrators dumped a pitcher of ice water upon his head, yelling "Wilson, you're all wet!" Now, fortunately, the revolution is over, and after a practical switch to the label "evolutionary psychology," professors of human behavior can now teach sociobiology on college campuses without inciting similar hostility.

Still, it is with some caution that I come to this, the most controversial chapter in the book, and note the seeming triviality that the same effects of the evolutionary biogeographical principles that apply to so many plants and animals are also readily apparent in humans. Gene flow among certain regional populations has, at times, been limited enough that people began to differentiate, slightly, and adapt to a great variety of climes and habitats. As with killer whales, variations in culture have also helped increase genetic diversity between populations. The various peoples of the Earth do show regional differences that should be embraced—rather than denied.

Why would such a trifling fact be controversial? The problem is that all political schemes, whether socialistic or bourgeois, democratic or elitist, must stand upon some theory of human nature. Typically, to have a strong view about governmental policies is to have a strong view about what it means to be human. Thus, politics has had a large and detracting influence in studies of human evolution and human biogeography just as it has with psychology, sociology, genetics, etc. Researchers attempting to negotiate these highly politicized fields have had to sail through the narrow passage between the Scylla of right-wing chauvinists and the Charybdis of left-wing radical biological-egalitarians, that is, between those who have tried to use Darwinism to promote their particular view of racial or ethnic superiority and those who have tried to support their political schemes with the fallacious premise that all people are born the same and so infinitely malleable and responsive to various social programs. As Richard Dawkins wrote in *The Ancestor's Tale*, "There are other projects for the study of human genetic diversity itself, which, bizarrely, come under recurrent political attack as though it were somehow improper to admit that humans vary. Thank goodness we do, if not very much."[1]

The following famous illustration of geno-graphical variation in human beings not only helps confirm Dawkins's point, it highlights the fact that accepting such truths can be quite beneficial. The example involves J. B. S. Haldane, who in attempting to use biogeography to solve a curious problem in human evolution, may have started a path for the possible cure for one of the most dreaded diseases still extant.

The mystery involved the sickle cell allele, a particular form of a gene that appears to control the shape of the red blood cell. Normal red blood cells tend to be slippery and doughnut shaped, perfectly suited to glide swiftly through the blood vessels, but those with the problematic allele produce red blood cells that are stiffer and sticky and, as you might expect, shaped like a sickle. Still, people born with an allele for each type of cell, normal and sickle cells, lead relatively healthy lives and have few adverse side-effects. But in the recessive case, when a child receives one sickle cell allele from each parent, the child will develop sickle cell anemia, an excruciating and frequently fatal disease caused by the myriad difficulties that arise from blood vessels getting blocked by too many sticky, clumping, sickle cells.

The question that Haldane pondered is why this allele for the sickle cell should be so common among certain populations, particularly those of African or Mediterranean ancestry. According to evolution, such frequency in these regions suggests that the allele must confer some sort of advantage, a survival benefit to those who carry the trait. Yet it seemed difficult to imagine what could be so beneficial to people with one copy of the allele that it manages to compensate for the devastating effect that it has on children born with two. Then Haldane compared the African and

Eurasian distribution of the sickle cell allele with that of malaria and found a conspicuous correspondence, each being prevalent in the same parts of West Africa, particularly south of the Sahara, the Middle East, India, and the Mediterranean, malaria having been quite prevalent in certain parts of southern Europe prior to 1920—before efforts to eradicate the mosquito. Studies in Ghana, Togo, and Greece in the 1960s and 1970s have indicated that this extraordinary correlation is even stronger and more precise on local levels—with the greatest frequency of the sickle cell allele typically occurring in towns most devastated by malaria.

As immediately surmised and as has now been confirmed, the sickle cell helps people fight off the bacterial blood disease malaria—a realization that may someday lead to a cure. Some researchers are now studying the way the sickle cell helps fight malaria in their quest for a vaccine. Such efforts underscore the importance of studying (and accepting) geodiversity within human populations.

Malaria has killed more people in the world than all other plagues and wars combined. Currently, it infests 200 to 500 million people worldwide each year, resulting in an annual death toll of more than 1 million, mostly children from sub-Saharan Africa. As Table 7.1 shows, the evolutionary computation behind the benefits of the sickle cell is as elementary as it is brutal.

Children with two normal red blood cell alleles will be vulnerable to malaria while children with both sickle cell alleles will develop sickle cell disease. Only those children who receive one copy of both will have protection from each illness. This is the vicious genetic lottery that many African children must endure. Evidently, malaria has been so murderous in certain regions that it was more genetically advantageous, in terms of producing the most successful offspring, for parents to have a sickle cell gene

Table 7.1.

Possible outcomes for children of parents with one normal and one sickle cell allele	Consequences for the child
Normal/normal	The child will be vulnerable to malaria.
Sickle/normal	The child will be healthy.
Normal/sickle	The child will be healthy.
Sickle/sickle	The child will have sickle cell disease.

and potentially have one quarter of their children suffer from sickle cell anemia rather than have all of their children potentially succumb to the deadliest consequences of this mosquito-borne illness.[2] It is hard to imagine an adaptation that seems less attributable to "intelligent"—or for that matter, compassionate—"design." And that is because the evolution and the spread of the sickle cell gene was not the result of careful planning by some omniscient entity but was simply the handiwork of that uncaring and remorseless mathematician—natural selection.

Although biogeography is a not a well-known science, everyone, at least in part, is a natural anthro-biogeographer. We can often determine the regional homeland of a person—or at least the regional homeland of many of his or her ancestors—just by glancing at certain facial features or skin tones—from the paler Northern Europeans or East Asians to the darker equatorial peoples.

The precise causes of the evolution of skin color have troubled researchers for a long time. Our earliest ancestors, being primates, almost certainly had light skin beneath dark hair—as this is typical of most primates and especially young chimpanzees, our closest relatives. Darker skin evolved in coordination with our loss of hair, which itself was likely an adaptation against disease-spreading, fur-loving parasites like fleas, ticks, and lice. And it is reasonable to surmise, as Darwin originally did, that darker skin had to evolve in the increasingly hairless people of the more sun-baked climes to protect them from the sun's rays.

But what was it about the sun's rays, specifically, that was so harmful? For decades, scientists believed that equatorial people evolved darker skin to give them protection against skin cancer, but skin cancer often occurs late in life—long after people have had time to pass on their light-skinned genes. Recently, Nina Jablonski and George Chaplin, another pair of married researchers, have determined that a more significant cause likely relates to two important nutrients, vitamin D and folic acid, which converts to folate in the body.[3] Intense sunlight reduces folate levels, which can lead to birth defects and can cause men to stop producing sperm. The melanin-darkened skin of tropical peoples can be seen as an adaptation that not only protected ancestors from sunburn but also helped ensure proper folate levels. The flip side to this adaptation is that sunlight also helps people manufacture vitamin D, which aids bone formation. Days are so long, and the sun is so strong in the lower latitudes that this still remains a significant source for the vitamin. People who had migrated into more northern climes evolved fairer skin in order to take full advantage of the lesser amount of sunlight.

The one well-known exception can be used to test the rule. The Inuit are a dark-skinned people who nevertheless inhabit the extremely sun-poor Arctic regions. Why? Their diets are unusually saturated with vitamin D. The skin of Arctic char; the blubber of seals, walruses, and whales; and the yolks of bird and fish eggs—all staples of the Inuit—are also all rich in vitamin D.

So the Inuit did not need sunlight to supplement their diets. Unfortunately, recent generations of Inuit, having incorporated a more westernized diet of mass-produced carbohydrates like breads, cereals, pastas, and pizzas, are experiencing an increase in rickets,[4] a disease brought about by vitamin D deficiency. Likewise, people of African descent who have moved to more northern climes are also especially vulnerable to rickets.

When did this evolution of whiter skin color, which has eventually led to so much geopolitical controversy, occur? According to the most recent molecular clock analyses, it happened within the past 20,000 years—a mere snap of the fingers in terms of evolutionary history. Also, blonde hair likely evolved in the past 11,000 years, and a recent study has suggested that blue eyes evolved from a single mutation in one ancestor within the last 6,000 to 10,000 years. All blue-eyed people on the planet are related through this lone and very recent common ancestor, and prior to this first blue-eyed individual, everyone had brown eyes. The movie "10,000 B.C." had, of course, a number of scientific errors, but it is amusing to note that any blue-eyed actor in the film would have been an anachronism.

As we will see throughout this chapter, other regional peculiarities have developed around the globe, some that are indeed more than skin or iris deep, but before we explore our great variety, there is something else about us that requires explaining, something that, given the extent of our distribution, is really quite extraordinary: our overall similarity.

This may seem an odd thing to state given some of the physical differences just mentioned. Yes, people from all over the world actually look very different, but in fact the genetic range of human beings is astonishingly narrow—despite appearances to the contrary. Remember that outward features can often obscure evolutionary relationships. The elephant shrew, which looks a lot like a shrew and very little like an elephant, is significantly closer to the latter. Brilliantly colored peacocks and the small, dull peahens are just different genders of the same bird. And black "panthers" and spotted leopards, which had been considered different species for centuries, are not only the same species, black panthers and spotted leopards can be born in the same litter. The black fur is simply a recessive trait—and black panthers today are now typically referred to by the less thrilling moniker, "melanistic leopards." The point is that superficial differences can hide genetic similarity, and this is especially the case in human beings. We are much more alike that we appear to be.

Some scientists, including Jared Diamond and Richard Dawkins, have suggested that the reason for our skin-deep variety is sexual preference. In other words, the superficial differences between the different peoples around the world have been intensified by local, cultural notions of attractiveness—which, perhaps, may have been instigated by the look of local rulers and by an inherent distrust or resentment of other peoples. That this explains some fraction of the differences is difficult to doubt. Various cultures of Western Europe have always had an obsession with, for lack of a better term, "whiteness," a fact so obvious, it scarcely needs documentation. In the "dark lady" sonnets of William Shakespeare, he notes that the addressee with her dark eyes and brows was changing traditional conceptions of beauty that favored light features:

> In the old age black was not counted fair,
> Or if it were it bore not beauty's name;
> But now is black beauty's successive heir,
> —William Shakespeare, "Sonnet CXXVII"

Even today, the term in English for being light-skinned is "fair", which is also a synonym for "pretty."

Other cultures also promote their own views of what is and is not desirable, based on the local phenotype. In the masked dance-dramas of Java, the refined and elegant characters, known as the *alus*, wear masks that exaggerate East Asian features. Typically, alusan masks have just a hint of a nose and very long, thin eyes. The coarser, stronger, and more common dancers, called the gagah, wear masks that amplify more Western facial characteristics with large noses and extremely round eyes. The gagah also play the demons. Such local, cultural differences in the conception of beauty may explain why people from various regions of the world look so different. Sexual selection may have exaggerated those variations that were skin-deep.

This is not to suggest that conceptions of beauty are exclusively a cultural by-product. Beauty is not a myth. Indeed, there are rather obvious evolutionary advantages to those qualities we find universally attractive—healthy skin, white set of full teeth, luxurious hair, the curves of a woman, the physical fitness of a man. All are visual clues to a healthy individual who is likely to produce healthy offspring. Thus, the adoring of these features transcends nearly all cultures and all times. Yet those particular notions of loveliness that change from place to place are certainly the result of the varying systems of those local and institutionalized idiosyncrasies that we call "culture," and this may have had an effect on how we look today.

Cultural barriers also explain why certain genetically similar populations are maintained even when they exist in close proximity and in relative harmony with other groups. The United States, populated as it is now by so many descendants of recent immigrants, provides myriad examples. American families of

East Asian, Indian, or Jewish ancestry not only practice differentiating traditions but tend to discourage marriage outside their respective groups. As with killer whale populations discussed in the last chapter, these learned tendencies help maintain an observable genetic structure despite the lack of geographical barriers.

Some may counter that physical differences also exist beneath the epidermis, and while that is true (consider the sickle cell gene above), these are almost always just differences in frequencies. Some groups exhibit certain traits more or less frequently than other groups—yet, interestingly, the extent of the variation within each group is always very close to the same. But when it comes to discussing human populations, even such seemingly innocuous statements still spark controversy.

As a case in point, in a 1995 speech before the British Association for the Advancement of Science, Sir Roger Bannister, the first man to run a sub-4-minute mile and a distinguished neuroscientist, said that he was willing to "risk political incorrectness by drawing attention to the seemingly obvious but under-stressed fact that black sprinters and black athletes in general all seem to have certain natural anatomical advantages." He then noted the "relative lack of subcutaneous fatty insulating tissue in the skin" in people of African ancestry, and other differences that might possibly explain all the gold medals. These comments generated an intense storm of denunciation.

The fear of course is that such statements may reanimate finally dead stereotypes that have been used in the past for abominable injustices. Harry Edwards at the University of California-Berkeley criticized the notion linking biology to athletic prowess thus: "What really is being said in a kind of underhanded way, is that blacks are closer to beasts and animals in terms of their genetic and physical and anatomical make up than they are to the rest of humanity. And that's where the indignity comes in."[5] But the notion that proficiency at running or jumping is an atavistic characteristic is perhaps one of the more unusual myths of which I am aware regarding human evolution, and it is particularly remarkable that this legend has persisted to such an extent given that one can instantly falsify it by noting that human are the only purely, non-hopping bipedal mammal. In other words, *two-legged proficiency is an advanced, derived, and peculiarly human trait—not a primitive one.* It was our less intelligent, ape-like ancestors who were struggling to get about on two feet, often using hands for help. It was the knuckle draggers who were slow. Over the course of millions of years, our ancestors steadily evolved to become more and more adept at walking and running, a process that required significant adaptations of the lower spine, pelvis, hip structure, hind-limb muscles, femora, fibula, feet, and toes. The classic drawing of the evolution of humans, showing the stages between a crouched, hairy form to the less hairy, upright man may also be subtitled "The evolution of bipedal speed."

The slowness of our ancestors has recently been emphasized by an analysis of the remains of small human-like "hobbits," with grapefruit-sized heads that lived on the island of Flores up until 13,000 years ago—what just may be the most astounding anthropological find in decades. While some scientists contend the bones were from a diseased group of pygmies, some with abnormally small skulls due to microcephaly, most other anthropologists who have published on the issue have now accepted that the group does represent a new hominid species, *Homo floresiensis*, a likely descendant of the East Asian *Homo erectus* tribes or perhaps of the smaller Dmanisi hominids of Georgia. If they are correct, then a primitive hominid cousin of the human race managed to coexist with modern human beings until nearly the dawn of civilization.

Recently, William Jungers of Stony Brook University analyzed the feet of *H. floresiensis* and discovered that they were hobbit-like as well—large and flat. He determined that the little people would have had a high-stepping gate and would have made extremely poor runners. He believes the feet of the Flores hominids are similar to those of ancestors from 2 to 3 million years ago, implying that our earlier ancestors were also very slow. So speed is one of the features that sets *Homo sapiens* apart, and those of us who are the most deftly bipedal—the agilest and the fastest—are actually flaunting a trait that is distinctly human. Or, to put it another way, arguing that two-legged speed is in any way less-than-human would be quite as silly as bats denouncing flying expertise as being less bat-like.

Pale skin, being the common skin color of many primates, is another atavistic trait—so slower, hairier, and paler people are, at least in this sense, far more like early hominids than others who are darker, have less body hair, and are faster. I bring this up not to suggest that "whites" are in some way more feral but to show that not only was the racist view of speed incorrect factually, its basic premise was flawed. We are all animals and are either more or less like our prehistoric ancestors in a great number of ways—having differentiated from them in some traits more than we have in others. But we not only still possess a number of obvious primate features, we still have a great number of primitive mammalian qualities—and even a great number of reptilian features. So sharing traits with primitive mammals is not necessarily insulting—and two-legged efficiency is not a trait we share with them, anyway.

The problem is that awkward efforts to explain away the domination of African descendants in certain Olympic and professional sports may exacerbate the situation by appearing to support implicitly the truly racist belief that there is something wrong with being innately fast. But the greatest weapon against prejudicial ignorance is more truth—not less.

Jon Entine provides an unblinking look at the issue in his *Taboo: Why Black Athletes Dominate Sport and Why We're Frightened to Talk about It*. As Entine points out, as of the year 2000, all 32 finalists of the 100-meter sprint from the previous four Olympics were of African descent—more specifically they were of *West* African descent, especially involving the coastal states Senegal, Nigeria, Cameroon, and Namibia. In fact, as of 2000, the fastest time for the 100-meter dash by a white person was 10 seconds, which ranks below 200 on the all time list.

Yet West Africans tend to be under-represented in the endurance sports. In the middle and longer distances, however, another African region, Kenya, has once again supplied an inordinate number of record holders. Indeed, the story of Kenyan runners may be even more surprising than the West African domination of the sprints. Between 1964 and 2000, Kenyans won 38 Olympic medals, including 13 golds in men's running events. Only the United States has more—and that is simply because of its high population of West African descendants. But, as Entine observed, what is truly remarkable is the geographical precision that bounds the Kenyan talent. The medal winners do not come from all parts of Kenya, which, like Iraq or the former (pre-1991) Yugoslavia, is an ethnically complex nation comprising a number of diverse groups. The swiftest lopers instead are concentrated among the pastoral Kalenjin people of the western highlands of the Great Rift Valley. Seventy percent of Kenyan's Olympic medals are distributed among the Kalenjin.

Remarkably, we can even narrow this down further. It is a particular group of the Kalenjin, the Nandi people who are cattle herders of the lush green hills of these highlands, who really stand out. Although they are merely one of at least nine ethnic groups comprising the Kalenjin, they have produced half of its star athletes. By 2001, runners from the Nandi district, roughly 500,000 people in total, had won 20% of the major international distance events. "By almost any measure," wrote Entine about the Nandi, "this tiny region in west-central Kenya represents the greatest concentration of raw athletic talent in the history of sports."[6]

Unfortunately, concentrating on events like the Olympics when determining ethnic and racial differences can be somewhat misleading. To focus on the Olympics is to concentrate on the exceedingly thin film on the top layer of humanity—the top one-hundred-millionth of a more thoroughly mixed genetic soup. School systems around the world are ever on the search for the fastest runners, resulting in a global vetting process that reaches into even the remotest towns. The fastest are continuously encouraged to run and are pushed into larger arenas. What results is an extremely precise selective process that skims only the fastest of the fastest, but this obscures the mixture of talents found among the different people of the world.

It may seem shocking that the top 200 speeds in the 100-meter dash have all been achieved by people of African descent, but such distinctions are actually expected from an evolutionary perspective. As shown throughout the other chapters, differentiation is the rule for such geographically extended groups, and few other non-flying terrestrial vertebrates live in so many places without having diversified into several species, without having acquired extremely specialized regional characteristics, both physical and behavioral. What is truly shocking is the pervasive lack of genetic diversity among humans. Consider Darwin's finches, which though confined to the Galápagos Islands, have split into thirteen different recognizable species, thus fractionalizing to a far greater extent than the whole, globally distributed human race.

Thus, we really should be perplexed that descendants of Europeans and Asians can even compete at all—and that many of them start appearing among the top thousand sprint times. As of 2008, the world record for the men's 100 meters was 9.69 seconds, held by Usain Bolt. The record among European whites is 10 flat. That is merely 31/100ths of a second difference over a distance of 100 meters, equating to an average speed of 22.74 mph (37.15 km/h) for the fastest West African descendant, and 22 mph (36 km/h) for the fastest man of European descent. Given the varying velocities of mammals from the pokey sloth to the cheetah (67 mph; 110 km/h), this is ridiculously close. It shows that the ranges of speeds of both groups are not that different. Moreover, in 2004, the gold medal winner at the Athens Olympics for the 400-meter sprint was Jeremy Wariner, a white man from Texas. In other words, for the 400-meter sprint, the range of possible outcomes for people of European and African descent was essentially identical. The difference is just in frequencies—not ranges. A greater percentage of people of West African descent are able to reach the highest speeds than that of European descent. Thus, the Olympics and professional American sports, while clearly underscoring a difference in frequencies of certain physical traits within ancestral populations that should no longer be denied, also gives a false impression of extreme genetic stratification that belies a general human physical similarity that transcends race.

In reality, it is the lack of genetic disparity between regional populations that is one of the more remarkable things about our species. Recent studies have shown that the biological differences among individuals within races accounts for 93 to 95% of genetic variation, while differences among the major races account for merely 3 to 5% of these variations. Richard Dawkins provided a rather clear way to understand these percentages: if everyone on the Earth were to disappear leaving only one local population—say, in sub-Saharan Africa or Central Europe or East Asia—then much of the spectrum of human biological variation would still be preserved. And much of what would be lost would be superficial—like, perhaps, blonde hair or the Asian epicanthal fold on the inner corner of the eye.

Another fact about human variation is just as telling. Chimpanzees, which reside along a relatively thin band straddling the African equator from Senegal in the west to Northeastern Tanzania, exhibit significantly more genetic range than do human beings, whose ancestors have been spread all throughout Africa and Eurasia for more than 70,000 (and perhaps even more than 1 million) years. In fact, any two chimpanzees chosen at random are likely to show, on average, twice as much genetic difference as any two randomly chosen human beings. We are a surprisingly uniform species.

In order to have a full understanding for the causes of this general similarity, we first need to discuss the biogeographical history of the human race. Fossil evidence suggests that H. erectus, one of our more recent hominid ancestors, evolved in Africa, and that some members of H. erectus migrated into Eurasia about 1.7 million years ago, reaching as far away as southeast Asia.

Essentially all anthropologists agree on that, but the question of what happened next has inspired two opposing views that have dominated the subject of human biogeography. The most popular view, labeled the "out of Africa-and-replacement" theory, suggested that modern humans evolved from H. *erectus* only in Africa, and then all descendants of the human race migrated into Eurasia within the past 100,000 years, "replacing" (read: exterminated and out-competed) all the other local hominid populations—rather than interbreeding with them. The alternative view, labeled the "multiregional theory," holds that modern human beings evolved everywhere—with European, Asian, and African hominids giving rise to European, Asian, and African H. *sapiens*. According to this view, genes would frequently continue to flow among the various populations to ensure parallel evolution throughout Africa and Eurasia and preventing speciation.

One perhaps can see that each of these two views could be alluring to some people due to their general geopolitical persuasion. The multiregional theory might seem more agreeable to more parochial-minded researchers from Eurasia, particularly to some Asian anthropologists who could prefer this theory because it aggrandizes the significance of East Asian H. *erectus* fossils and turns China into a cradle of Eastern humanity. Meanwhile, the out-of-Africa view may be preferable to others for exactly the opposite reason, appealing more to anti-parochial, egalitarian-minded Afrocentrists. This is not to suggest that either of these views is in any way illegitimate or unscientific, for both seemed to enjoy support from certain data.

For a while, researchers were concluding that they had confirmed a rigid recent-out-of-Africa-with-replacement hypothesis, again and again, while falsifying the multiregional view.

Perhaps, the most well-known discovery came from a DNA analysis in the 1980s that implied everyone on Earth is a descendant of an ancestral and African "mitochondrial Eve" who, if we are to accept molecular clock rates, lived about 140,000 years ago. We know this ancestor was a female because mitochondria are passed exclusively down through the mother. We believe she was African because the lineages that branched earliest from the mitochondrial tree are all found in Africa. Similarly, another DNA analysis of more than 12,000 East Asian men, including Australian aborigines, concluded that that the Y-chromosome Adam lived in Africa less than 100,000 years ago. Many researchers used these studies to conclude that all descendants of the human race migrated out of Africa within the past 100,000 years.

While these examples are indeed evidence for the out-of-Africa view, they do not provide as strong a proof as some suggest. The problem is that each of these studies was not pinpointing the ancestral homeland of a single people, who then took over the world; each was simply determining the ancestral homeland of a single inherited trait. The spread of such a trait from a certain region no more requires the uniform motion of a single population that supplants all other people than does the spread of a single idea or a single disease. These three human-carried entities—genes, ideas, and diseases—can be relayed geographically from person to person, town to town, and region to region, among already existent populations.

Consider the precise dissemination of the sickle cell gene described at the beginning of this chapter. Clearly, such a distribution does not demand that a single population of sickle cell carriers began in one particular sickle cell Eden, then migrated all throughout those regions in Africa and Europe, taking up residence everywhere, without interbreeding with local populations. Instead, many of the ancestors of the people across Africa and Eurasia who share the sickle cell trait were already in place, and the gene then became infused throughout these populations as people from neighboring towns and regions continued to meet and couple. When the gene reached a town ravaged by malaria, it would prove helpful and so increase in frequency among its inhabitants. Whenever it came to a region not affected by malaria, it began to decrease in frequency. Indeed, we could imagine a world in which every region was so plagued by malaria that the gene eventually became universal. In such a malaria-infested, sickle cell world, all humans would then be descended from some "sickle cell Adam" (or Eve), but this would not mean that the birthplace of sickle cell Adam was also the birthplace of humanity.

The same is true for mitochondrial Eve. Yes, we are all her descendants, but she is not *the* one and only greatest of great-great-great-etc. grandmothers; she is just one of our many such grandmothers. She is one of the important contributors to our genetic makeup, our one common female ancestor who is exclusively on our maternal side—our mother's, mother's, mother's, mother's, etc., mother. But we also have countless other ancestors who were alive at that time, many of whom contributed other genes to the current human race. And some of these people may have lived somewhere else. So it is important to pay careful attention to her name or you may be misled. Mitochondrial Eve is not the mother of the human race; she is the mother of our particular mitochondria.

There has always been another issue with the strict out-of-Africa-with-replacement theory, which may be underscored by a comment made by a friend of mine at a far-too-serious cocktail party discussion on the topic. As a few people were debating whether our more *sapiens*-like ancestors that left Africa ever interbred with the Eurasian *H. erectus*, my slightly inebriated friend asked, "Well, wait a second. When was beer invented?" Although everyone laughed, the larger point is that odd mating decisions probably preceded beer. And, given the way people are today, it perhaps does seem unlikely that our ancestors would have been dissuaded by a large brow ridge. Still, it is, of course, possible that interbreeding was so infrequent that the resulting offspring left no descendants alive today.

One recent analysis involving multiple DNA regions by Alan Templeton has helped merge the out-of-Africa and multiregional views.[7] According to his research, the vast majority of the genes that make us human did come from Africa, but many Eurasians (and even some Africans) also have genes that have flowed down from Eurasian *H. erectus*. He concluded that there were three major migrations of our ancestors from Africa into Eurasia. The first, by *H. erectus*, occurred 1.9 to 1.7 million years ago—and as noted in the multiregional theory, genes still continued to flow among all of these widely separated populations on the two continents. This original scattering of *H. erectus* preceded two later expansions of primitive African "Empires," a pre-*sapiens* movement occurring about 700,000 years ago and another 100,000 years ago. This wide-scale motion of African clans into the old world would not have been a coordinated takeover, of course, but likely comprised a series of successful invasions and migrations of various tribes throughout the old world. These African clans may have slaughtered many of the local *H. erectus* tribes, but they also settled down and mixed with their populations— the very type of thing that we find so frequently in recorded human history. Each time these invading Africans brought a little more humanness with them, much the way the Romans brought roads.

Templeton's conclusions helped blend the two theories—and might be best described as the out-of-Africa again and again and again (with interbreeding between all the regions) theory. Much of our humanness did evolve in Africa, but many of us still have some genes from those local Eurasian populations of H. erectus.

Why did so many evolutionary advances occur in Africa? Well, it helped that Africa probably boasted larger populations and a greater amount of genetic variety among the hominids, so individuals there had a better chance for successful adaptations. Some of these developments clearly gave some particular African group significant advantages over the rest of the hominid world, including improved weapons, communication, and increasing abilities to plan and strategize, leading to more advanced techniques in hunting and warfare. This allowed them to migrate, survive, out-compete, and conquer—all facilitating the spread of their genes and the success of their descendants. According to Templeton's analysis, the second movement from Africa 700,000 years ago resulted in the expansion of the "Acheulean culture,"[8] a primitive tribal system that made use of sharpened, pear-shaped stones that were used like hand-axes. The last movement from Africa, 100,000 years ago, seems to have brought smaller brow ridges, higher vertical foreheads, rounder skulls, and more prominent chins to Eurasia, all indicative of greater intelligence and vocal skills.

This view of human origins would be consistent with the lack of genetic diversity among humans. Templeton's paper noted that despite the extraordinary geographical range of our ancestors—encompassing rainforests, deserts, mountains, and wide rivers—genes have continued to course among the populations without any significant interruption for the last 1.5 million years. Thus, in this view, we may imagine these steady streams of genes between all regional populations being flooded by the two large waves of genes washing out from Africa and inundating Eurasia 700,000 and 100,000 years ago.

The primary hominid feature that allowed us to maintain this gene flow—allowed us to spread the humanness, so to speak—was a combination of an increasingly daring and venturesome nature, matched with a remarkable physical mobility. We often tend to look at our species as somewhat slow, frail, and vulnerable—kept alive only by our craftiness. The horse and antelope, after all, are faster, the otter and beaver are better swimmers, the hyena has more endurance, the chimpanzee is a better climber. But each of these animals is a specialist in each of these particular endeavors. Not only do we have our own athletic forte—our manual dexterity allowing us to throw and manipulate objects with astonishing facility[9]—but we are also fairly good at all the other aforementioned mammalian specialties. We are better swimmers than the horse, antelope, hyena, and chimpanzee—indeed almost all animals that are not semi-aquatic. Likewise, we can climb better than the vast majority of all non-arboreal animals. And practically everything that can either climb (chimpanzees, lemurs, squirrels) or swim (beavers, otters, seals) better than humans, we can easily outrun for distances greater than 100 meters. We seem almost to have evolved to negotiate long-distance obstacle courses made up of natural Earthly barriers. If we were to include a number of these events in an interspecies decathlon—such as diving off a cliff into water, swimming across a mile-wide river, lunging deep to touch the bottom of the river, climbing to the highest part of a tree, climbing to the highest parts of a mountain, running for over 25 miles (40 km)—a human would very likely take the gold, with, perhaps, the leopard taking the silver. The fact is, human beings are athletic generalists of the first rank, and few other vertebrates even come close.

This is one of the key characteristics that helped maintain our extraordinary equality. Chimpanzees, by contrast, tend to be confined to forested or near-tree locales and are almost completely unable to swim. Thus, the various rivers that run through their thin stretch of African habitat—and particularly the Congo—have divided them into groups, accelerating their genetic diversity. We remain unbounded by rivers or straits or forests or deserts or even mountains—and we are the only placental mammal, other than rats, to colonize Australia within the past 40 million years, having reached the isolated continent at least 50,000 years ago, even before the invention of modern water-craft. Our tremendous terrestrial mobility, our wandering instincts, and our daring in the face of seemingly daunting boundaries and exotic new territories have all ensured that no place has remained genetically isolated for too long. And this has managed to limit differences and create an equality among human beings that we do not observe in other creatures of similar geographical extent.

Other than regional variations in our physical appearances, something else has also appeared to conflict with the narrowness of our genetic differences—and that is the great chasm in technological progress that has separated the different peoples of the Earth. This was the issue that Jared Diamond confronted in *Guns, Germs, and Steel*. As Diamond asked: "Why did wealth and power become distributed as they now are, rather than in some other way? For instance, why weren't Native Americans, Africans, and

Aboriginal Australians the ones who decimated, subjugated, or exterminated Europeans and Asians?"[10]

The answer is not easy—and for a long time, many European scientists accepted the simplest explanation: innate differences in intelligence and inventiveness gave the Europeans a technological advantage. Many leading intellectuals and academics in the last half of the twentieth century rejected that belief, denouncing it as racist. Yet they were never able to advance a more persuasive reason. Recently, however, Jared Diamond has used his considerable knowledge of biogeography to solve what in essence was a biogeography-of-technology problem.

First, according to Diamond, the Eurasians got an insurmountable head start because, 10,000 years ago, their continent contained significantly more plants and animals that were naturally predisposed to cultivation—that were, in essence, agricultural by nature. At that time, the Fertile Crescent, which describes the lush crescent-shaped Middle Eastern highlands that ran between the Red Sea, Mediterranean Sea, and Persian Gulf (through what today is Egypt, Israel, Turkey, Syria, Iraq, and Iran) was in proximity to a number of tasty, tamable, easy-breeding, and sustentative animals and boasted many nourishing plants. Wheat, barley, and peas—all of which provide a lot of protein and are particularly productive in their wild state—grew naturally in these regions. Hunter-gatherer societies of the area started settling down around these natural fields, developing villages based, at least partly, on a diet of wild wheat, barley, and peas. These plants are particularly advantageous because they are self-pollinators, grow quickly, are easy to harvest, are readily stored, and are predisposed to being farmed. Indeed, the only difference between our farm wheat and its wild ancestors is a single mutation. The seeds of wild wheat and barley grow at the top of a stalk that spontaneously shatters, dropping the seeds to the ground. But in the more agriculturally desirable "mutant" form, the stalk does not shatter—allowing one to easily collect the seeds. It was this mutant stalk that was most often available for these Ur-farmers to pick, who would then plant their mutant seeds. This led to wheat and barley that became increasingly easy to harvest—a sort of accidental agriculture.

Residents of the Fertile Crescent were also fortunate to live in a region that boasted four of the five wild mammals that would become the "major five" farm mammals—all of which resided in west or central Asia:

The Asiatic mouflon sheep of west and central Asia—the wild ancestor to all the world's sheep;

The bezoar goat of west Asia—the wild ancestor to the domesticated goat;

The extinct aurochs of Eurasia and north Africa—the wild ancestor to cow, cattle, and alias ox;

The wild boars of Eurasia and north Africa—the wild ancestor to the domesticated pigs;

The extinct wild horses of southern Russia.

Not all animals are equally easy to domesticate, and Jared Diamond shows that, in general, the plants and animals in other regions were not so accommodating. The most valuable farm animals, for example, need to grow fat and contented on cheap and readily available plants, and they have to have a social hierarchy, be easily controllable, have gentle dispositions, and cannot be prone to panic. Hippos and buffalo are too dangerous, zebras too nasty.

Also, analogous to the fortuitous cultivation of wheat, barley, and peas, it is probable that animal husbandry was another accidental science. Imagine that some group of hunters had managed to capture a large group of Asiatic mouflon sheep—too many for their tribe to eat all at one time. It would have been wasteful to kill all of them, thus surrendering many to the flies. Recall again the similar conservational strategy practiced by killer whales, which often nudge seals to safety when they are not hungry and are through using them in their aggressive hunting games. This conservational strategy, which appears natural to these most intelligent mammals, suggests a possible behavioral precursor to husbandry. The idea that the hunters should not only let the superfluous sheep live but also try to keep them trapped until they were hungry again would not have presented a major cognitive leap.

The prolonged plenty afforded by a large capture would have led the tribe to try to recreate this successful hunt, that is, to attempt to trap a great number of sheep at one time and keep as many confined and alive for as long as possible. Thus, these early sheep hunters were now starting to become the first shepherds. Moreover, as with the first domesticated crops, it is not that significant a genetic step to get cooperative farm sheep from the relatively docile Asiatic mouflon sheep—and it seems plausible that this process was also inadvertent. The first shepherds did not try to breed a more docile species of sheep, but, once having them penned, they probably soon realized it was best to kill the belligerent and flighty ones first.

Each of these animals—sheep, cow, goat, and pig—was likely domesticated in a different part of the Fertile Crescent and then the livestock started to spread throughout the rest of this agricultural Eden. Not too far away, the people of China also had the good fortune of living in a place conducive to agricultural discovery and were developing similar techniques. Thus, it was in these blessed places of Eurasia where life became easiest and where extra-culinary matters could be afforded some attention. Here is where civilization first started to flower.

Within a few thousand years, the crops, animals, and agricultural discoveries of the Fertile Crescent easily expanded along similar climate zones and geography, which is to say, throughout the east–west plane of Eurasia. But this organic foundation to civilization could not spread through the prohibitive climes of equatorial Africa or across the frozen Bering bridge or for that matter across the straits of Wallacea and into Australia. Communication, trading, and most importantly, the exchange of ideas also continued to flow east and west through the easily traveled Eurasia. This was not the case in Africa and the Americas, continents that are elongated mostly north–south, resulting in a great variety of climatic zones that restricted or prevented the flow of information among their populations. Thus, the people of North America and the southern realms were not only effectively cut off from all the important discoveries occurring in Eurasia, they were also often isolated from the advances produced by nearby tribes. Once again, we find that the current distribution of continents, brought about by that volcanic ring around Antarctica and that served as an evolutionary catalyst among northern hemisphere mammals, has also had an extraordinary impact on human civilizations, particularly in helping forge such a massive technological divide.

In *Guns, Germs, and Steel*, Jared Diamond uses the principles of biogeography to explain the current distribution of wealth and technology with a particular focus on the question of why people of Eurasia have developed so many more helpful and powerful inventions than peoples from other regions. Recall again the reasons why Laurasian mammals have managed to dominate the world as discussed in Chapter 5. The geographical connectedness of Eurasia and North America (and Africa for a while) provided a multi-continent platform for evolutionary potential. Competition was ferocious, and any advance, like the development of the mammalian Carnivora, spread throughout the Laurasian regions. These, in turn, became subject to more evolutionary improvements. A few successful carnivore ancestors led to the more successful canine and feline lines—which in turn produced even more sophisticated and efficient survivors, the gray wolf and the leopard.

The same thing occurred with technologies. For the past 10,000 years, Eurasia has been a continent that allowed much easier communication of ideas and technologies. Any helpful invention or important modification of an existing technology, whether developed in China, England, or anywhere in between, spread all throughout Eurasia. And this in turn would enable anyone living anywhere along the giant continent to adopt the new invention and tinker, modify, and improve. The relatively unobstructed layout of Eurasia that promoted the evolution of the fiercest and most efficient mammals also provided an environment that, for the same reasons, promoted the evolution of the fiercest and most efficient technologies. Once again a group of northerners had the advantage—while the isolated clans in other parts of the world—specifically North America and the Gondwanan regions—New Guinea, Australia, New Zealand, Southern Africa, South America—were at their mercy.

In brief, isolation slows the pace of evolutionary progress, whether organic or technological. This is why many of the

THE BATTLE OVER EDEN

Gondwanan continents not only have so many vulnerable "living fossils," plants and animals that still conform to ancient and primitive designs—but also had similarly vulnerable people who were dependent on ancient and primitive technologies. As with the tuatara so with the culture and inventions of the Maori—the isolation of New Zealand managed to preserve the seemingly archaic nature of both. Any helpful adaptations occurring anywhere else in the world could not reach these regions and out-compete the primitive forms. And those few advances that we do find in New Zealand had to occur to those comparatively few denizens who inhabited these isolated islands.

Diamond's biogeography-based explanation for the difference in technological progress among the various peoples of the world is difficult to deny—and certainly far more persuasive as a primary cause than the early-twentieth-century genetic one. Indeed, it would be somewhat unreflective for Eurasians to associate the use of stone-age technologies with inherent mental narrowness—as their own ancestors, who were genetically identical to modern Eurasians, employed stone-age technology for tens of thousands of years and only developed more advanced tools relatively recently. Certainly, people with similar inherent talents and dispositions as Newton, Edison, Huygens, or Darwin existed during this dark and vulnerable era, but they would have rarely had the time or the inclination to work their talents on anything other than the short-term survival of their family and themselves. More, even when they had those rare moments of leisure and reflection, they could only apply their ingenuity and insight to the primitive beliefs and tools that surrounded them. Any novel ideas that they may have developed, they could have only communicated verbally to those nearby. The same may be said of the indigenous people of New Guinea, Australia, and South America.

The morning of October 12, 1492, is a seminal moment in human history—not because this is when people discovered America, for as school children are taught, Native Americans had entered the Americas more than 13,000 years ago. Nor is it important because it represents the first time Eurasians managed to cross an ocean to reach the New World. Indeed, as we shall see in the last chapter, Polynesians had managed to conquer the giant and turbulent Pacific, reaching the coast of South America more than 400 years prior to the European crossing of the narrower Atlantic. Indeed, Columbus and his crew were not even the first Western Europeans to reach the New World, for it appears that Norse voyagers lived in settlements in Newfoundland in AD 1000. The significance of this day in 1492, the day when Columbus's ships first reached the Bahamas, is that it represents the colossal historical moment when the biogeographical gap between West and East closed for good.

In Chapter 1, I discussed the *Ensatina* salamanders of California as an example of a ring species that surrounds the San Joaquin Valley. Ring species describe a series of closely related neighboring populations that form a large geographical circle, eventually bringing together two "end populations" that are so genetically distinct they no longer interbreed and can be classified as two species. Columbus's landing in the Bahamas represents a similar and permanent closure of a biogeographical circle of human beings. Although this ring did not involve speciation, it is still one that flows around a genetic continuum that brings together the two most distantly related peoples together again. Consider the steady changes in human populations as you begin in Western Europe and head east into Central Europe and then into Russia and Mongolia—as the frequency of blue eyes, blondes, and red-heads continues to dwindle, and skin tone continues to change almost imperceptibly from paler tones to ochres and reds. The Native Americans that reached the East Coast of North America represents a group that is most distantly related to the Western Europeans, the final end point of the ring. And they must have been stunned when first seeing those monstrous ships and the strange human beings that inhabited them—the paler skin, the increased facial hair. Imagine how alien the blondes and redheads must have seemed.

Because of the eventual westward expansion of Europeans across North America, the biogeographical ring no longer forms a neat circle ending in the East Coast of North America, but now overlaps, with the boundary of Westerners reaching into Alaska. There we still find a genetic divide, somewhat maintained by cultural differences, between the Inuit and the Alaskan descendants of Europeans.

Columbus's finding of the "New World," filled with previously unknown inhabitants, was a great source of European pride, a milestone in west-focused history, leading to the colonization of North America and its eventual growth into a superpower. Native Americans, of course, do not share this celebratory point of view, and over the latter half of the twentieth century, many Americans of European descent have been urged to consider Columbus Day from the viewpoint of the original inhabitants. From this perspective, it is a particularly tragic moment in history, one that precipitated the devastation of the Native American peoples and, some would say, rampant ecological destruction.

Yet, it is also important to remember that the most deadly thing that the Europeans carried with them was neither their guns nor their steel (swords), but their germs. For the same biogeographical principles that led to the most advanced weapons also produced the deadliest diseases. Measles, smallpox, tuberculosis, and 'flu had all spread to Eurasians through farm animals, eventually adapting to their new hosts. Quickly these diseases proliferated throughout the Eurasian populations due to their high population densities and easy communication. In response, Eurasians evolved too. People with natural antibodies started to proliferate among Europeans—as did strategies of quarantining, which helped limit the spread of the diseases. But isolated populations like those in the New World had not developed any immunity and did not know enough to be cautious around the sick. Smallpox was particularly devastating to the Aztec. Within less than 100 years, from 1520 to 1618, Mexico went from a population of 20 million to about 1.6 million. Similarly, when Columbus reached Hispaniola in 1492 the native population was 8 million. By 1535, no one was left alive.

If we try for a moment to look at the subject scientifically, we can see this European entry into the New World as the realization of two biogeographical events—the closing of a genetic ring (as previously discussed) and the colonization of a previously isolated realm by invasive populations, both humans and germs, now finally able to cross a particular barrier. Just as the rising of the Isthmus of Panama closed the Caribbean gap between the biota of North and South America, so Columbus's famous voyage had now helped close the Atlantic, bringing a population to

North America that had benefitted from thousands of years of an accelerated, massive-continental evolution of technology and disease resistance. Just as the northern taxa wreaked havoc on the South American aboriginals—and just as the introduction of rats or dogs to islands decimates many of the endemics—so too did European populations out-compete and kill the Native American populations.

We tend to vilify Western Europeans as the ones who are the most ethno-chauvinistic, most genocidal, most destructive of the ecosystem, and this is not an incomprehensible position. When we consider all the brutal conflicts inflicted upon the Earth by people of European descent, the plight of the Native Americans, the Holocaust of World War II, slavery, colonialism, the Crusades, and their pollution spewing industrialization—much of their activity has had bloody and disastrous consequences. But all these rather infamous achievements in war and pollution are really more a function of technological progress rather than due to some inherent predisposition to belligerence unique to Europeans. The xenophobia, tribalism, seeming paranoia, and hatred of the other—all these are not Euro-American inventions; they are very human characteristics. Indeed, they are more appropriately described as mammalian.

Territory and resources are indispensible to survival—and many animals make it a point to kill competitors. Wolves kill coyotes at every opportunity; and lions slaughter cheetahs with similar enthusiasm. But the greatest hostility is often saved for members of the same species, who provide the toughest competition for food, territory, and mates. Chimps and wolves are violently territorial and frequently attack new and unknown members of their own species who have wandered into their territory. The same instinct is conspicuous in humans, as we see with the innumerable clans discovered in New Guinea throughout the twentieth century, who, without any imperialistic persuasion were quite practiced at tribal warfare. In Columbus's October 12th journal entry, the same day he first reached the West Indies, he noted that many of the friendly new-found natives had scars on their bodies, which they had indicated were the result of battles with other nearby islanders or perhaps the Native American mainlanders who had tried to take them as slaves. Studies have shown that a far greater percentage of our early ancestors died during intertribal battles than did people during the blood-soaked twentieth century.[11]

Historians, politicians, and religious leaders often put forth many complicated religious–political–historical and economic reasons for the causes of ethnic conflict. But an impartial alien biologist, new to this world and studying animal and human behavior, would almost certainly scoff at such convoluted rationalizations. The alien would see intraspecies warfare among chimpanzees, wolves, and humans—and ascribe to it the same sociobiological cause. Humans are xenophobic by nature and rationalize their hatreds or suspicions of others through a combination of facts and myths, logic and sophistry. It seems that the easiest thing to get a person to believe is that the nearby people who look slightly different and have a lot of land are really evil and do not deserve their comfort or their riches: They are heretics and pray to the wrong God; or they are savages and less than human; or they are decadent and do not merit their wealth; or they have taken it from us in the past and we should get it now; or they are planning the same thing and are about to kill us (these last two rationalizations may often be true). One does not need to be a great proselytizer to urge people to battle or convince them they truly deserve what other people have.

The intense suspicion and irrational hatred for another nearby ethnic group was not invented in 1930s Germany or by American settlers of the west but is an attitude that transcends all cultures and all times—and has also been prevalent in Africa, Asia, and the pre-colonial New World. The sinister view of a seemingly monolithically evil and ruthless white people slaughtering peaceful, environmentally friendly Africans or Native Americans is just as much a simplistic and vilifying stereotype as the racist propaganda used by Europeans and Americans to rationalize their bloody conquests. The unfortunate truth is that the technological progress of Europeans—their guns and their steel—better equipped them to bring to fruition their less altruistic urges. They weren't really any more desirous of conquest than poorer societies; they were just more efficient at it.

This, of course, is not an effort to excuse the brutalities mentioned. We are civilized enough now, on the whole, to overcome and rise above these baser mammalian urges. So while we can understand their origin, we can never condone their practice. Many nations in the twentieth century finally made an attempt to reverse their land-stealing, genocidal tendencies—or at the very least recognize the injustice and evil of it, rather than rationalize it away with abstract theories of "manifest destiny," "master race," "agrarian reform," or a "worker's paradise." The very basic biogeographical causes that led descendants of Western Europeans to develop so many helpful technologies, enjoy such artistic success, and establish the foundations of the major scientific disciplines are the same ones that allowed them unparalleled success in subjugation and warfare. Modern medicine, Beethoven's Ninth, Newton's *Principia*, and Darwin's *On the Origin of Species* are on one side of the scale while slavery, genocide, and imperialism are on the other. These landmarks in European history comprise some of the best and worst that our species has to offer—the by-products of basic human qualities all magnified by the same biogeographical imperatives that have been operating on the mammals of Eurasia for tens of millions of years.

CHAPTER 8

The Grand Coalescence of Life and Earth

The past, present, and future of biogeography

"Here Be Dragons" is a phrase that most people associate with the outskirts of ancient, sepia maps—the ominous warning scrawled over those places where the lands leave the familiar. At least, that is where many people think we would find the phrase. As noted in the Preface, "Here Be Dragons" is unknown from historical maps, appearing only once, in Latin, on the Hunt-Lenox Globe, constructed in the years after Columbus's trip to the New World. The phrase was not, as so many believe, a general warning to sailors about alien realms. It was, instead, one of the first recorded post-Columbian biogeographical remarks and has now become, perhaps, the most famous distributional comment ever, likely marking the general region where tales of the Komodo dragon originated.

We find many other similar pronouncements on medieval documents. The *Tabula Peutingeriana*, an atlas of the road system running through the Roman Empire from Britain to India, redrawn from more ancient charts in 1265 by a Colmar monk (in modern France), has *in his locis elephanti nascuntur* ("in these places elephants are born"), and *in his locis scorpiones nascuntur* ("in these places scorpions are born"). Similarly, the Cotton Tiberius Map from 1025 has *hic abundant leones* ("Here lions abound"). These examples underscore the prevalence of the biogeographical impulse of medieval cartographers to locate and communicate the origins of wondrous creatures.

The map of antiquity that may be of the most interest to students of organic distributions is the *Carta Marina* map (1539) of Scandinavia and surrounding seas, drawn by the exiled Swede Olaus Magnus in Venice. The long subtitle of the map underlined its intent to include the "marvels" or "peculiar...wonders of nature" of the northern lands. This massive and colorful document, measuring more than 3 feet (1 m) wide and nearly 7 feet (2 m) tall, is busy to the point of bursting with pictorials of Scandinavian culture and wildlife—men hunting with arrows, men ice fishing, people praying, men on skis and skates, a woman milking a reindeer, and a plethora of northern animals, some imagined, some real. The drawings, often unrealistic in scale, are a somewhat innocent and awkward mixture of reality and mythology—and have the primitive and earthy charm common to antique charts.

Magnus's depictions of reindeer, foxes, wolves, boars, snakes, pelicans, and cranes show some accuracy, including that of an eagle catching a white hare, suggesting creatures familiar to the artist, but some of the more exotic creatures, and particularly the sea life, were archaic guesses. In the Atlantic Ocean, Magnus drew whales like giant, hideous, toothy fish with water jetting from fleshy pipes on their heads—and one of his whales is being attacked by a smaller version that he labels "orcha." Magnus also filled the sea with monsters of all sorts, including a large red serpent attacking a ship and a giant lobster grasping a man in one of its claws. Imagine a jumbled, medieval graphic novel, forced all into one frame, and you probably have a good idea of Magnus's work.

Perhaps the most interesting drawings were his efforts to portray the very northern animals, like the seals on ice floes being hunted by men with harpoons in the Bay of Bothnia, a crude depiction of what may be a walrus at the northern tip of Norway, and a very accurate depiction of polar bears in Iceland. One of the polar bears is sitting on an ice floe just off the coast of the Atlantic isle with a fish in its mouth. While Iceland is 186 miles (300 km) from Greenland, polar bears have often ended up in Iceland when ice floes have carried them near the island.

These faunal picture-stories on Magnus's *Carta Marina*, like *hic abundant leones* or "Here be dragons," denote some of the first crude guesses of people interested in organic distributions. They help capture the basic desire of these geographers to discover and communicate what came from where. Eventually, their nineteenth-century counterparts would add much light to their dark gropings and the resulting knowledge would then be used to change our view of the world.

The study of biogeography has progressed greatly since the days of Magnus, and the development of new analytical tools has proved indispensable to the organic researcher. Molecular analyses of DNA have allowed researchers to sort out relationships previously deemed baffling, often confirming the following tenet: "biogeography typically trumps taxonomy and anticipates molecular phylogeny." That is, among closely related groups, geographical distance is often a better predictor of relationships than is morphological similarity because natural selection can encourage parallel forms in distantly related creatures. Marsupial wolves (now extinct) and marsupial moles resemble their placental counterparts far more closely than they do each other, but these appearances are deceiving. The distributional fact that nearly all the pouched animals reside exclusively in Australasia is truly the informative clue—implying that marsupial wolves and moles (and kangaroos and koalas) are all more closely related to each other than they are to any geographically distant placental which they may superficially resemble. The extinct pygmy mammoths of Channel Islands, just outside of California, and the extinct pygmy mammoths of Wrangel Island, north of Siberia, present a similar case. After a quick inspection, it might seem that the two pygmy forms would be sister species. But biogeography helps clarify: the pygmy mammoths of Channel Islands descended from Californian populations of the Columbian mammoths (*Mammuthus columbi*) while the Wrangel Island pygmy mammoths derived from Siberian populations of the tundra mammoth (*Mammuthus primigenius*).

One of the most intriguing and useful developments in biogeography in recent years was the discovery that protein can survive intact in fossils for tens of millions of years and may now be used to sort out confusing evolutionary relationships among long extinct taxa. Previously, the ephemeral nature of DNA, which degrades within tens of thousands of years, had limited molecular analyses to the currently living or recently extinct. But in 2007, intact protein was extracted from the collagen of a *Tyrannosaurus Rex* femur, more than 68 million years old, and subsequent comparisons to protein in the bone collagen of chickens helped confirm that modern birds are indeed descended from dinosaurs. This technique greatly extends the scope of such analyses, allowing us more detailed views of the distributional relationships of fossil taxa.

Perhaps, the most exciting new idea in biogeography will be its continued application to the study of ideas, technologies, and other human artifacts—a trail pioneered by Jared Diamond in *Guns, Germs, and Steel*. This would expand the subject's already prodigious scope—and bring many seemingly disparate elements into a single unifying scheme.

E. O. Wilson's work *Consilience* clarifies and espouses the principle of scientific materialism that all phenomena, even such multifaceted abstractions as the arts and the humanities, are reducible through a strict series of causes and effects to a few principles in physics. Everything is linked through physical, mechanistic processes. Everything is interconnected—sociology, biology, chemistry, physics. When considering the place of biogeography within this orderly and grand estate of universal happenings, it may not be unreasonable to say that it comprises what may be, from our point of view, the largest and most significant wing, encompassing the surface of the Earth and all its gaudy and dramatic occurrences.

Astronomers train their lens on distant specks in the sky; the elements of chemistry exist at the microscopic, and those of physics are smaller yet. But biogeography is the unifying science that describes the familiar, touchable universe that exists between the realms of the far and the small. It describes the bodily world around you now, the world you can hear, feel, see, and smell—the green grasses, the stately shade trees, the flutters of movement of the furry and the feathered. The entire animated landscape that envelops you at this moment in all its sensual variety—all of it reduces to the material principles of evolution interacting with the chaotic events and pressures of the Earth's surface.

Biogeography thus reminds us that the word "origin" denotes both a process *and a place*—that the great variety of life did not just arise in some indistinct and misty nowhere. Instead, location matters. When we study distributions we begin to associate the evolution of plants and animals with a particular setting, thus providing a tangible background to the birth and development of species. More, we discover that the environment was not just a passive backdrop, completely disassociated from the evolutionary modifications occurring within it, but was an active causal participant, supplying the mechanistic forces driving those changes. The Earth is not merely the cradle of life; it is its womb.

Indeed, we need not even limit ourselves to the realm of the purely organic, for, as we have seen, the principles of biogeography relate not just to plants and animals but to all of their constructs. Researchers, for example, have now begun applying the principles of biogeography to the mud-nesting habits of swallows and martins.[1] As they discovered, the distribution of certain nesting habits throughout the New World and Africa parallel the biogeographical history of the birds. We can also do the same for our own species, tracing the past migrations and genetic flow of populations by tracing the distributions of languages, architecture, cuisine, etc. Thus, these simple distributional tenets apply broadly to your surroundings whether you are in your landscaped backyard filled with a variety of store-bought plants, or at the beach listening to seagulls and sipping an iced coffee, or in Manhattan surrounded by skyscrapers, taxis, and pigeons. *Everything*—and *every component and trait of everything*—has both an evolutionary history and a place of origin. And this simple fact not only has extraordinarily revealing consequences, it is the most direct path to a complete understanding of the observable world about you.

In order to emphasize the full illuminating power of biogeography, let us assume that, right now, you are sitting at a table on the beach at sunset in Maui about to indulge in a Hawaiian luau. Everything on your plate is traditional. The eponymous dish, the luau, consists of chicken baked in coconut milk, mixed with juvenile taro leaves. You also have sweet potatoes, slices of mountain apples, poi, which is a paste-like staple of Hawaii made from taro roots, and some slivers of pig that had been roasted in an underground pit. For desert you have sweet mixtures of pineapples, coconut, and bananas. And, in order to heighten the mood, you are wearing a traditional lei made of leaves and blossoms from Hawaii's state tree, the Kukui tree.

As always, a biogeographical understanding brings clarity. Although all of this fauna and flora on your plate and around your neck would seem so authentic—imbued with the true flavor and aroma of the island you are visiting—in reality, none of it is native to Hawaii. And except for the pineapple, which was brought to Hawaii from South America by Captain Cook around 1770 and first cultivated by European immigrants, all of it was brought to the tropical islands by Polynesians. In brief, what most people think of as naturally "Hawaiian" was really part of a portable, seagoing, Polynesian ecosystem—and if we trace the biogeographical history of all these elements we can discover many secrets of this venturesome society: their origins and the extent of their Pacific conquests.

The Polynesians were ancient seafarers from Taiwan, the first Pacific pioneers. They sailed on giant double-hulled canoes with planks fastened between them to hold people and cargo, navigating by stars and bird migrations. Their first jaunts from Southeast Asia were merely tens of miles, likely to islands that they could see. Eventually, they traveled to and past New Guinea, reaching the Solomon Islands by 1600 BC, then to Vanatua. From there, the marine gaps increased to hundreds of miles, and they became the first people to cross them, settling on Fiji, Tonga, and Samoa by 900 BC. Still, they pressed further with even more daring voyages eastward into the middle of the Pacific, reaching the Cook Islands, Bora Bora, Tahiti, and the Marquesas island groups. Then, from the middle of the great Polynesian triangle, they embarked on their greatest journeys yet, now at times more than 2000 miles over open ocean, to its three exotic and far-flung corners: Hawaii to the North (AD 400–500), Easter Island to the East (perhaps as late as AD 1200), and New Zealand to the southwest (~AD 1200). Their route to the latter is particularly surprising. Imagine the amazement of anthropologists when they discovered that the Maori of New Zealand came not from the continents of Australia or Southeast Asia, but from a small population on Tahiti, far to the east in the middle of the Pacific.

It is likely that, prior to colonization of each of these island groups, the Polynesian societies sent smaller, exploratory expeditions in search of new islands to conquer. Once the voyagers returned with tales of discovery, a much larger faction of hopeful colonizers may have then been gathered. Like many migrating peoples, the Polynesians continued to bring helpful elements of their past homeland with them, particularly the edible crops and animals.

The genetic trail of the pig, the centerpiece of the luau, follows the path of the Polynesians and can be traced back through the islands of West Polynesia through Timor, Fores, Java, Sumatra, and back to Vietnam.[2] Polynesians also brought the taro from Southeast Asia, the *wauke* (or paper mulberry) from South China, and the mountain apple and Kukui tree, which provided the blossoms for the lei, from Malaysia. Bananas, coconuts, and the domestic chicken are also all East Asian imports, spread around the Pacific islands by these sun-baked gods of the sea. True enough, coconuts float and show some tolerance to saltwater, allowing them to colonize nearby islands naturally. But their resistance to the sea is not unlimited, and studies suggest coconuts will not germinate after significant periods in the ocean. It was the Polynesians who carried the coconut to the remotest islands.

But there is one thing on your plate that did not come from East Asia—the sweet potato, which actually originated in South America. How did this tuber end up in Hawaii? For many decades, anthropologists assumed that Spanish explorers dispensed the sweet potato on the Pacific Isles during the exercise of their colonial ambitions, but archaeological discoveries have revealed the sweet potato was already on Cook Island by AD 1000–1100 and on Hawaii prior to AD 1430, long before the Europeans had ventured so deep into the Pacific. Is it possible that Polynesians actually reached the Americas before Columbus?

Linguistic evidence would seem to support pre-Columbian contact. In the Quechua language of native Peruvians and Ecuadorians, the word for sweet potato is *cumal*—very similar to the Polynesian word, *kumala*.[3] But even more significant evidence was discovered in 2007 when a radiocarbon and DNA analysis on chicken bones recovered in Chile ended all doubts about the extraordinary extent of Polynesian travels. According to anthropologist Alice A. Storey and colleagues, the chicken bones date from 1304 to 1424 and are genetically identical to chicken bones found in pre-historic Polynesian sites.[4] Incredibly, these seafaring people had conquered the vast and turbulent Pacific long before Europeans had even managed to cross the narrower Atlantic.

Thus, we can get a hazy idea of this first contact between those intrepid Polynesians and native South Americans. Given the early age of the Cook Island sweet potatoes, an exploratory seafaring group had to have reached the Western Coast of South America prior to AD 1100, which in turn would suggest they may have traveled from Marquesas, not from Easter Island, which would not be settled until AD 1200. Their interaction with the aboriginals must have been friendly. Though working with very different languages, they were able to communicate their agricultural technologies, trading the chickens and their methods of raising them for the sweet potatoes and their secrets. It is also likely that the Polynesians introduced the native South Americans to coconuts at this time, thus explaining the presence of the coconut tree in the Americas. Genetic analyses show that there was no interbreeding between Polynesians and native South Americans. So once these voyagers returned to their home islands with their new found sweet potatoes—and tales of the giant landmass and their fascinating inhabitants—the Polynesians evidently never returned to colonize the continent, perhaps because it was already inhabited or, perhaps, because a large group that tried became lost at sea.

The question of who first crossed an ocean to reach the Americas has been a fiercely debated issue for decades, yet this answer, along with much of the Polynesian history, can be revealed through simple biogeographical knowledge of the typical items served at a luau. The fact that none of the traditional meal was truly indigenous to Hawaii also underscores the meagerness and rugged hostility of the world in its natural state.

Hawaii prior to its discovery by Polynesians—with its lack of domesticated crops and lack of non-avian vertebrates—was far from Elysian and would have proved a difficult place to survive. Instead, what many people really imagine when they think of a "natural tropical Pacific paradise" is really a carefully modified environment, one that has been artificially transformed to meet Polynesian needs, both culinary and aesthetic.

Throughout the course of this book, I have attempted to show the power of distributional evidence. The material and unbreakable bonds of heredity ensure that all life is connected, that all sister species must be physically linked through some recent ancestor that moved about the Earth according to its physical abilities. Recognition of this simple fact has revolutionized our views of Earth, life, and human history. Indeed, given certain problematic disjunctions between flora and fauna on either side of the Pacific, currently the basis of debate, some biogeographers have even suggested that we need to change our theories of the tectonic history of the Pacific or even overhaul conventional views of planetary science. So biogeography may be at the forefront of yet another major scientific revolution. Regardless, the 2007 analysis of ancient Chilean chicken bones, which confirmed pre-Columbian Polynesian contact with South America, shows that the capacity for biogeography to solve mysteries remains unabated.

Imagine life evolving on a planet covered by a single, uniform ocean—of constant depth, stable temperature, and few currents, and you have imagined a planet where life would very likely remain simple and relatively homogenous. It is the constant unrest along the Earth's surface, the ever-changing landscapes, that helps provide barriers and create a variety of selective pressures. The geological and climatic diversity of this planet is what promotes the diversity of its species. But, as we have seen, despite the wondrousness of the results, the transformations have often been the result of rather harsh and unforgiving processes. Many of those same features that make our planet such an efficient engine of speciation also often make it a severe and disastrous place to live.

The common refrain of this book, enunciated so clearly by the biogeographer Croizat, is that life and Earth evolve together. But it is really less a partnership than a battle. All of the special endemic flora and fauna of Galápagos and Hawaii will eventually be lost to the sea, and the only survivors will be descendants of those that manage to reach the newer islands. In Antarctica, the extraordinary cold-weather adaptations of the emperor penguins came at the expense of the slow freezing or starving of all other year-round vertebrates. We have seen something similar on every continent and in all oceans; the changing environments continually stress the local residents, forcing them to adapt or perish. Even the current global diversity of mammals, it is important to remember, arose out of the carcasses of dinosaurs. The Earth's breathtaking biodiversity, so wondrous in one sense, can also be seen as the combat scars lining the mantle of living tissue that envelops its outer shell, telling of life's constant battle to survive the shifting features of this volatile and unreliable planet.

But while the study of evolution, at bottom, may be a study of deterministic mercilessness, governed by simple and dispassionate principles, the biogeographical history of this planet has provided the most voluptuous and gorgeous drama in the known universe. We are often told that sciences are cold and inert while only the arts are beautiful and provocative, but this neglects the compelling nature of the history of the Earth's surface. It is, in essence, an enormous series of acts and scenes that are, at once, harsh, bloody, exhilarating, tragic, and marvelous—*Hamlet* writ large. Those made privy to the secrets of biogeography have enhanced their innate appreciation of biodiversity by discovering what it truly is—a spectacular by-product of the often fiery and cataclysmic forces that have sculpted the planet's surface. Earth and life evolve together, and the end result of this ceaseless co-evolution has been a series of striking biogeographical patterns that have transformed our essential views of who we are and where we live.

In this book, we have mostly dwelled on major Earthly upheavals—the breakup of Gondwana, the emergence of the Isthmus of Panama—that have had the greatest effects upon life. But other barriers form at regional levels, leading to the smaller scale patterns of diversity that we see on a daily basis. What is true globally is also true locally. Rivers, forests, and grasslands form narrower, ephemeral boundaries that still manage to isolate and alter. Speciation follows and life continues to become more splendid.

Every day our backyards and town parks, the nearby forests and lakes, offer a library of life that too few people know how to read and enjoy. Even those fascinated by ecology and the natural world often remain innocent of biogeography, but this subject is far too important to remain the obscure and treasured diversion of geniuses. As we have seen throughout this book, all plants and animals are part of the same vast system of genetic tributaries, as material streams of genes have all flowed without interruption over the surface of the globe from the same ancestral source. This one simple idea has led experts in the distributions of plants and animals to revolutionize biology, geology, and anthropology—and this, in turn, should inspire us to adopt the same expansive perspective. For once we can see the global biosphere as they did, once we truly understand the beauty and power of this grand coalescence of life and Earth, we will never look at the world in the same way again.

ENDNOTES

PREFACE: "THAT GRAND SUBJECT"

1. In *A Devil's Chaplain* (New York: Houghton-Mifflin, 2003), Richard Dawkins writes that "Daniel Dennett has credited Darwin with the greatest idea ever to occur to a human mind" (66), perhaps also expressing a view Dawkins shares. Dennett's quote is from *Darwin's Dangerous Idea* (New York: Simon and Schuster, 1995).

CHAPTER 1: GALÁPAGAN EPIPHANY

1. *The Autobiography of Charles Darwin*, ed. Nora Barlow (1887; New York: Barnes and Noble, 2005), 18. All quotes from Darwin's autobiography come from this edition.

2. Darwin would write in his autobiography that he expected it would surprise many people to discover how religious he had been, having remained "orthodox" even during the first years of his voyage. As Darwin wrote: "Whilst on board the *Beagle* I was quite orthodox, and I remember being heartily laughed at by several of the officers (though themselves orthodox) for quoting the Bible as an unanswerable authority on some point of morality" (*Autobiography*, 66). Still, even as a young man, Darwin had met with other views that almost certainly made him more amenable to the unconventional discoveries that awaited. He came from a family of freethinkers, and his grandfather, Erasmus, had suggested in his writings that species may have come from simpler forms. Darwin also was familiar with Larmarck's ideas on the mutability of species, and he brought on his voyage a copy of Lyell's *Principles of Geology*, which described important evidence for the ancient history of the Earth.

3. Fortunately for scientific historians, we need not speculate on the observations that motivated Darwin because no scientist has ever been more detailed or forthcoming—in journal entries, in private letters, in his autobiography—about the conceptual development of an idea. We can

follow the emergence of the theory of evolution from its roots in his "Journal of Researches," kept during his five-year trip around the world. Or we can retrace it through his autobiographical reminiscences, written near the end of his life, about his progression from a young natural theologian, likely destined for a quiet existence at a county parish, into the most controversial scientist of the Victorian era. It would not be overreaching to say that we know more about the conceptual evolution of evolution—about the origin of *Origin*—than we do about the development of any other scientific viewpoint. And we know from Darwin's own statements on the matter precisely what facts compelled him, what observations confirmed in his mind the vast interconnectedness of life. As Darwin makes clear, time and again, it was biogeography, above all else, that led him to the theory of evolution.

In his autobiography, Darwin details some of those observations:

> During the voyage of the *Beagle* I had been deeply impressed by discovering in the Pampean formation great fossil animals covered with armour like that on the existing armadillos; secondly, by the manner in which closely allied animals replace one another in proceeding southwards over the Continent; and thirdly by the South American character of most of the productions of the Galápagos archipelago, and more especially by the manner in which they differ slightly on each island of the group; none of these islands appearing to be very ancient in a geological sense.

> It was evident that facts such as these, as well as many others, could be explained on the supposition that species gradually become modified; and the subject haunted me. (44)

4. Charles Darwin, *The Voyage of the Beagle, in From So Simple a Beginning; The Four Great Books by Charles Darwin*, ed. E. O. Wilson (New York: W. W. Norton, 2006), 341. All quotes from *The Voyage of the Beagle* and *On the Origin of Species* will come from this edition.
5. *The Voyage of the Beagle*, 341.
6. *On the Origin of Species*, 702–3.
7. For example, the hairs of a polar bear's white fur are hollow and filled with air providing both insulation and buoyancy. It is a common, though false, scientific legend that the hollow hairs act like fiber optic wires and reflect sunlight toward their black skin, which then absorbs it. Instead, the polar bear hair is peculiarly well suited to keep heat from radiating

outward—not for trying to transmit it inward. Beneath the skin is yet another layer of insulation, thick layers of fat.

8. *On the Origin of Species*, 700.

9. A few Creationists of today try to explain island distributions by resorting to Noah's Ark. Such animals could not reach these islands after the Flood. But the tale of the Flood leads to even more biogeographical problems. Imagine each one of the 13 different species of Darwin's finches flying from the Ark through Eurasia into North America, then into South America, and then to Galápagos, without any individuals of the group ever ending up anywhere else. Now imagine the same problem with all 140 species of Australian marsupials—from the marsupial mouse to the Kangaroo.

10. The distribution of these spiders extends to so many remote regions it has led to what may be the only known biogeographical joke: As Neal Armstrong once said, "One small step for man…, hey, is that a long-jawed orb-weaver?"

11. Recent phylogenetic analyses have shown that Darwin's finches are most closely related to another group of birds, only two species of which reside in South America, and the rest live in the Caribbean. This invites the possibility that the Galápagan finches dispersed directly from the Caribbean, but it seems that the most likely explanation for this distribution is that both the Caribbean and Galápagan finches all descend from the same South American ancestor.

12. Ring species highlight how evolutionary processes can occasionally conflict with our desire to divide taxa into well-defined groups we call "species"—a difficulty that scientists now refer to as the "species problem." For example, researchers currently classify the *Ensatina* salamanders into seven different "subspecies."

13. *On the Origin of Speices*, 461.

14. Ibid., 468.

CHAPTER 2: THE *MESOSAURUS* PROBLEM

1. James Gleick, *Genius: The Life and Science of Richard Feynman* (New York: Pantheon Books, 1992), 63.

2. I have a vague feeling that someone else may have written something very close to this—that he or she would give an alien a copy of *On the Origin of Species* as the greatest achievement of the human race—but, if so, I cannot remember who. If someone has, then I apologize for not

giving proper attribution, but I do sincerely share the same sentiment. Also, it is possible that before I handed the alien *On the Origin of Species*, I would pick up Lucretius' *De Rerum Natura* first, pause for a moment, then finally decide on Darwin's masterpiece.

3. A. R. Wallace, "On the Law which Has Regulated the Introduction of New Species," *Annals and Magazine of Natural History* (1855), Available at: http://www.victorianweb.org/science/science_texts/wallace_law.html

4. Chamberlin's quote appears in A. Hallam, *Great Geological Controversies* (Oxford: Oxford University Press, 1990), 152.

5. G. G. Simpson, "Antarctica as a Faunal Migration Route," *Proceedings of the Sixth Pacific Science Congress of the Pacific Science Association* 2(1940), 755–78. Available at; http://www.wku.edu/~smithch/biogeog/SIMP940A.htm; *idem*, "Mammals and the Nature of Continents," *American Journal of Science* **241**(1943), 1–31. Fortunately, both Simpson's papers and du Toit's rejoinder are currently accessible at "Early Classics in Biogeography, Distribution, and Diversity Studies: To 1950," http://www.wku.edu/~smithch/biogeog/—a biogeographically oriented website maintained by Charles H. Smith. The book by Alexander du Toit that Simpson often referenced was *Our Wandering Continents; An Hypothesis of Continental Drifting* (Edinburgh/London: Oliver and Boyd, 1937).

6. Simpson, "Antarctica as a Faunal Migration Route," 756–7.

7. Alexander du Toit, "Tertiary Mammals and Continental Drift. A Rejoinder to George G. Simpson," *American Journal of Science* **242**(1944), 145–63, esp. 147.

8. Or it is possible you could imagine a range slightly less wide by assuming a track into Northern Europe and into Greenland, then southward into South America.

9. The dearth of mammalian species or genera that actually have such a cosmopolitan distribution—from South Africa to South America—shows how incredible such a scenario is.

10. Du Toit, "Tertiary Mammals and Continental Drift," 151.

11. Simpson, "Mammals and the Nature of Continents," 2.

CHAPTER 3: PYGMY MAMMOTHS
AND MYSTERIOUS ISLANDS

1. Charles Darwin, *Charles Darwin's Diary of the Voyage of the H. M. S. Beagle*, ed. Nora Barlow (Cambridge: Cambridge University Press, 1934), available at http://www.galapagos.to/TEXTS/DIARY.HTM

ENDNOTES

2. R. Werner and K. A. Hoernle, "New Volcanological and Volatile Data Confirm the Hypothesis for the Continuous Existence of Galápagos Islands for the Past 17 m.y.," *Int J Earth Sci,* **92**/6(2003), 904–11.

3. Charles Darwin, *The Voyage of the Beagle, in From So Simple a Beginning; The Four Great Books by Charles Darwin,* ed. E. O. Wilson (New York: W. W. Norton, 2006), 345.

4. Ibid., 330–1.

5. Ibid., 329.

6. C. C. Austin and G. R. Zug, "Molecular and Morphological Evolution in the South-Central Pacific Skink *Emoia tongana* (Reptilia: Squamata): Uniformity and Human-Mediated Dispersal," *Australian Journal of Zoology* **47**(1999), 425–37.

7. Donald Lee Johnson, "Problems in the Land Vertebrate Zoogeography of Certain Islands and the Swimming Powers of Elephants," *Journal of Biogeography* **7**/4(1980), 383–98, esp. 385.

8. Ibid., 383.

9. Chris Lavers, *Why Elephants Have Big Ears: Understanding Patterns of Life on Earth* (New York: St. Martin's Press, 2000), 23–7.

10. Bryan G. Fry et al., "Early evolution of the venom system in lizards and snakes," *Nature,* **439**(2006), 584–8; doi:10.1038/nature04328.

11. Jared M. Diamond, "Did Komodo Dragons Evolve to Eat Pygmy Elephants?", *Nature* **326**(1987), 832.

CHAPTER 4: THE VOLCANIC RING
THAT CHANGED THE WORLD

1. See Dennis McCarthy, "Geophysical Explanation for the Disparity in Spreading Rates between the Northern and Southern Hemispheres," *Journal of Geophysical Research* 112(2007), B03410, doi:10.1029/2006JB004535, for a more detailed analysis of the geometric consequences of the northward motion of continents away from Antarctica.

2. Donald R. Prothero, *The Eocene-Oligocene Transition: Paradise Lost* (New York: Columbia University Press, 1994), 16.

3. M. A. Reguero, S. A. Marenssi, and S. N. Santillana, "Antarctic Peninsula and South America (Patagonia) Paleogene Terrestrial Faunas and Environments: Biogeographic Relationships," *Palaeogeography-Palaeoclimatology-Palaeoecology* **179**/3–4(2002), 189–210.

ENDNOTES

4. The harsh wintering of emperor penguins can be seen in the stunning French documentary, *March of the Penguins* (2005), which received a well-deserved Academy Award for Best Documentary Feature.

5. William J. Murphy et al., "Resolution of the Early Placental Mammal Radiation Using Bayesian Phylogenetics," *Science* **294**/5550(2001), 2348; doi:10.1126/science.1067179. See also S. Blair Hedges' excellent discussion in "Afrotheria: Plate Tectonics Meets Genomics," *Proceedings of the National Academy of Sciences* **98**(2001), 1–2.

6. The recent study by J. Harshman et al. ("Phylogenomic Evidence for Multiple Losses of Flight in Ratite Birds," *Proceedings of the National Academy of Sciences* **105**/36(2008), 13462–7; doi:10.1073/pnas.0803242105), which placed the weakly flying tinamous within the ratite group, has now raised an interesting question about ratite evolution: Was the common ancestor to the ratite–tinamous group flightless, and did tinamous reacquire the ability to fly? Or did this common ancestor have the ability to fly—and did ostriches, rheas, and the ancestor of the Australasian ratites all independently become flightless as they evolved on different continents? Harshman et al. conclude that the latter scenario was more likely because there is no confirmed example of a previously grounded bird reacquiring the ability to fly. Moreover, that three different ratite ancestors would each lose the ability to fly is not as improbable as it seems. The tinamous is a fast-running, ground-dweller that only takes to the air when seriously threatened and then can only stay in flight for short distances. It seems likely that the most recent ancestor of the ratites and tinamous was also a poor-flying, ground dweller and that it had developed certain behavioral adaptations that predisposed many of its descendants to flightlessness.

7. A. Cooper et al., "Complete Mitochondrial Genome Sequences of Two Extinct Moas Clarify Ratite Evolution," *Nature* **409**(2001), 704–7.

8. Christopher J. Humphries and Lynne R. Parenti, *Cladistic Biogeography: Interpreting Patterns of Plant and Animal Distributions* (Oxford: Oxford University Press, 1999).

9. John S. Sparks and Wm. Leo Smith, "Freshwater Fishes, Dispersal Ability, and Nonevidence: 'Gondwana Life Rafts' to the Rescue," *Syst. Biol.* **54**/1(2005), 158–65, esp. 162.

ENDNOTES

CHAPTER 5: THE BLOODY FALL OF SOUTH AMERICA

1. Ministry for the Environment, *State of New Zealand's Environment 1997*, "The State of Our Frogs"; available at URL:http://www.mfe. govt.nz/publications/ser/ser1997/html/chapter9.7.4.html
2. D. Brawand, W. Wahli, and H. Kaessmann, "Loss of Egg Yolk Genes in Mammals and the Origin of Lactation and Placentation," *PLoS Biology* **6**/3(2008), e63; doi:10.1371/journal.pbio.0060063.
3. As quoted by Larua Spinney in "The Lion King," *The Guardian* (July 24, 2003); available online at URL:http://www.guardian.co.uk/science/ 2003/ jul/24/science.research. See also S. Wroe, "A Review of Terrestrial Mammalian and Reptilian Carnivore Ecology in Australian Fossil Faunas," *Australian Journal of Zoology* **50**(2002), 1–24.
4. This helps underscore the rationale behind another famous quip of J. B. S. Haldane: When asked by a theologian whether his biological studies had led him to form any new insights about the creator, Haldane replied, "He has an inordinate fondness for beetles." All told, we have discovered 450,000 species of beetles around the globe, and each year we discover 2,400 more. Some researchers estimate that there may be as many as 8 million species, vastly outnumbering the 9,000 bird species and 4,000 mammal species.
5. Jared Diamond, *The Third Chimpanzee: The Evolution and Future of the Human Animal* (New York: Harper Perennial, 2006), 225. All of the facts discussed about New Guinea highlanders come from this book.
6. Ibid., 229.
7. Christopher W. Dick, Eldredge Bermingham, Maristerra R. Lemes, and Rogerio Gribel, "Extreme Long-Distance Dispersal of the Lowland Tropical Rainforest Tree *Ceiba pentandra* L. (Malvaceae) in Africa and the Neotropics," *Molecular Ecology* **16**/14(2007), 3039–49; doi:10.1111/j.1365–294X.2007.03341.x

CHAPTER 6: ENCHANTED WATERS

1. Zofia A. Kaliszewska et al., "Population Histories of Right Whales (Cetacea: Eubalaena) Inferred from Mitochondrial Sequence Diversities and Divergences of their Whale Lice (Amphipoda: Cyamus)," *Molecular Ecology* **14**/11(2005), 3439–56. See also "Secrets of the Whale Riders: Crablike 'Whale Lice' Show How Endangered Cetaceans

Evolved," *ScienceDaily* (Sept. 14, 2005); retrieved April 9, 2008 from http://www.sciencedaily.com/releases/2005/09/050914104455.htm

2. J. D. Durand, A. Collet, S. Chow, B. Guinand, and P. Borsa, "Nuclear and Mitochondrial DNA Markers Indicate Unidirectional Gene Flow of Indo-Pacific to Atlantic Bigeye Tuna *(Thunnus obesus)* Populations, and Their Admixture off Southern Africa," *Mar. Biol.* **147**(2005), 313–22. See also Pilar Martínez, Elena G. González, Rita Castilho, and Rafael Zardoya, "Genetic Diversity and Historical Demography of Atlantic Bigeye Tuna *(Thunnus obesus)*," *Molecular Phylogenetics and Evolution* **39**/2(2006), 404–16.

3. See C. Mary Fowler and V. Tunnicliffe, "Influence of Sea-Floor Spreading on the Global Hydrothermal Vent Fauna," *Nature* **379**(1996), 531–3; and *eaedem*, "Hydrothermal Vent Communities and Plate Tectonics," *Endeavour* **21**/4(1997), 164–8.

4. Daniel R. Brooks, "Origins, Diversification, and Historical Structure of the Helminth Fauna Inhabiting Neotropical Freshwater Stingrays (Potamotrygonidae)," *Journal of Parasitology* **78**/4(1992), 588–95.

5. Russell W. Mapes, Afonso C. R. Nogueira, Drew S. Coleman, and Angela Maria Leguizamon Vega, "Evidence for a Continent Scale Drainage Inversion in the Amazon Basin Since the Late Cretaceous," Geological Society of America, *Abstracts with Programs* **38**/7(2006), 518.

6. Carol Kaesuk Yoon, "Parasites Hint Amazon River Once Flowed West," *New York Times* (May 18, 1993).

7. Chris Lavers, *Why Elephants have Big Ears: Understanding Patterns of Life and Earth* (New York: St. Martin's Press, 2000), 155–78.

8. Jukka U. Palo and Risto Vainola, "The Enigma of the Landlocked Baikal and Caspian Seals Addressed through Phylogeny of Phocine Mitochondrial Sequences," *Biological Journal of the Linnean Society*, **88**/1(2006), 61.

9. Lance G. Barrett-Lennard and Kathy A. Heise, "The Natural History and Ecology of Killer Whales," in James A Estes et al. (eds), *Whales, Whaling, and Ocean Ecosystems* (Berkeley: University of California Press, California, 2006), 169. Much of the information about killer whales discussed here was found in this splendid chapter of the book.

10. The CNN video of the killer whales using waves to hunt can be seen on YouTube: http://www.youtube.com/watch?v=rBF9cDBUakA& feature=related

11. A. Rus Hoelzel et al., "Evolution of Population Structure in a Highly Social Top Predator, the Killer Whale," *Molecular Biology and Evolution* **24**/6(2007), 1407–15.

ENDNOTES

CHAPTER 7: THE BATTLE OVER EDEN

1. Richard Dawkins, *The Ancestor's Tale* (Boston: Houghton Mifflin Harcourt, 2004), 39.
2. William H. Durham, *Coevolution: Genes, Culture, and Human Diversity* (Stanford: Stanford University Press, 1991), 124.
3. Nina G. Jablonski and George Chaplin, "The Evolution of Human Skin Coloration", *Journal of Human Evolution* **39**(2000), 57–106.
4. World Institute of Natural Health Sciences, "Changes in Arctic Diet Put Inuit at Risk for Rickets," June 10, 2007; available online at URL: http://www.winhs.org/news/news20070610.htm. Retrieved January 2008.
5. Quote appears in Jon Entine, *Taboo: Why Black Athletes Dominate Sports and Why We're Afraid to Talk about It* (New York: Public Affairs, 2001), 3.
6. Ibid., 40.
7. Alan Templeton, "Out of Africa Again and Again," *Nature* **416**(2002), 45–51.
8. "Acheulean" is derived from Saint-Acheul in northwestern France, which is one of the first places where anthropologists discovered many of these tools.
9. Many of us can throw a baseball more than 230 ft (70 m) with remarkable accuracy or repeatedly toss basketballs into hoops 9 ft (3 m) off the ground while standing 16 ft (5 m) away. Moreover, we can juggle, keeping three balls in the air, tossing and catching almost subconsciously. No animal of course can match humans in any such feats. The ability to throw (and catch) comes so easily that we are not even really impressed with it as a talent. But if any other animal could impel objects such distances with similar accuracy we would find it remarkable indeed.
10. Jared Diamond, *Guns, Germs, and Steel: The Fates of Human Societies* (New York: Norton, 1997), 15.
11. Jung-Kyoo Choi and Samuel Bowles, "The Coevolution of Parochial Altruism and War," *Science* **318**/5850(2007), 636–40.

CHAPTER 8: THE GRAND COALESCENCE OF LIFE AND EARTH

1. David W. Winkler and Frederick H. Sheldon, "Evolution of Nest Construction in Swallows (Hirundinidae): A Molecular Phylogenetic

ENDNOTES

Perspective," *Proceedings of the National Academy of Sciences of the United States of America* **90**/12(1993), 5705–7.

2. Greger Larson et al., "Phylogeny and Ancient DNA of *Sus* Provides Insights into Neolithic Expansion in Island Southeast Asia and Oceania," *Proceedings of the National Academy of Sciences* **104**(2007), 4834–9; doi:10.1073/pnas.0607753104

3. A. Montenegro, C. Avis, and A. J. Weaver, "Modelling the Pre-historic Arrival of the Sweet Potato in Polynesia," *Journal of Archaeological Science* **35**(2008), 355–67.

4. Alice A. Storey et al., "Radiocarbon and DNA Evidence for a Pre-Columbian Introduction of Polynesian Chickens to Chile," *Proceedings of the National Academy of Sciences of United States of America* **104**/25(2007), 10335–9, esp. 10335.

BIBLIOGRAPHY

Attenborough, David, *Life on Earth: A Natural History* (London: Collins/BBC, 1980).

—— *The Living Planet* (Boston: Little, Brown, 1984).

Barrett-Lennard, L. G. and K. A. Heise, "The Natural History and Ecology of Killer Whales," in James A. Estes et al. (eds), *Whales, Whaling, and Ocean Ecosystems* (Berkeley: University of California Press, 2006).

Carroll, R. L., *Vertebrate Paleontology and Evolution* (New York: W. H. Freeman, 1988).

Colbert, E. H., M. Morales, and Eli C. Minkoff, *Evolution of the Vertebrates: A History of the Backboned Animals through Time* (New York: Wiley, 1991).

Cowen, R., *History of Life*, 3rd edn (Boston: Blackwell Science, 2000).

Cox, A. and R. B. Hart, *Plate Tectonics: How It Works* (Oxford: Blackwell, 1986).

Darwin, C., *Journal of researches into the natural history and geology of the countries visited during the voyage of H. M. S. Beagle round the world* (London: John Murray, 1845).

—— *On the origin of species by means of natural selection, or the preservation of favoured races in the struggle for life* (London: John Murray, 1859).

—— *The Autobiography of Charles Darwin*, ed. Nora Barlow (1887; London: Collins, 1958).

—— *The Voyage of the Beagle*, in *From So Simple a Beginning; The Four Great Books by Charles Darwin*, ed. E. O. Wilson (New York: Norton, 2006).

Dawkins, R., *A Devil's Chaplain* (New York: Houghton-Mifflin, 2003).

—— *The Ancestor's Tale* (Boston: Houghton Mifflin Harcourt, 2004).

Dennett, D., *Darwin's Dangerous Idea* (New York: Simon and Schuster, 1995).

Diamond, J., *Guns, Germs, and Steel: The Fates of Human Societies* (New York: Norton, 1997).

—— *The Third Chimpanzee: The Evolution and Future of the Human Animal* (New York: Harper Perennial, 2006).

Durham, W. H., *Coevolution: Genes, Culture, and Human Diversity* (Stanford: Stanford University Press, 1991).

BIBLIOGRAPHY

du Toit, A. L., *Our Wandering Continents; An Hypothesis of Continental Drifting* (London: Oliver and Boyd, 1937).

Eastman, J., *The Book of Field and Roadside: Open-Country Weeds, Trees, and Wildflowers of Eastern North America* (Mechanicsburg, PA: Stackpole Books, 2003).

Ebach, M. C. and R. S. Tangney (eds), *Biogeography in a Changing World* (Boca Raton, FL: Taylor and Francis, CRC Press, 2006).

Entine, J., *Taboo: Why Black Athletes Dominate Sports and Why We're Afraid to Talk about It* (New York: Public Affairs, 2001).

Gleick, J., *Genius: The Life and Science of Richard Feynman* (New York: Pantheon Books, 1992).

Hallam, A., *Great Geological Controversies* (Oxford: Oxford University Press, 1990).

Humphries, C. J. and L. R. Parenti, *Cladistic Biogeography: Interpreting Patterns of Plant and Animal Distributions* (Oxford: Oxford University Press, 1999).

Huxley, J. S., *Evolution: The Modern Synthesis* (London: Allen and Unwin, 1942).

Kitcher, P., *Abusing Science: The Case Against Creationism* (Cambridge, MA: MIT Press, 1982).

Lavers, C., *Why Elephants have Big Ears: Understanding Patterns of Life and Earth* (New York: St. Martin's Press, 2000).

MacArthur, R. H. and E. O. Wilson, *The Theory of Island Biogeography* (1967; rpt Princeton, NJ: Princeton University Press, 2001).

Pollan, M., *The Botany of Desire: A Plant's-Eye View of the World* (New York: Random House, 2001).

Prothero, D. R., *The Eocene-Oligocene Transition: Paradise Lost* (New York: Columbia University Press, 1994).

Sagan, C. *The Demon-Haunted World* (New York: Random House, 1996).

Wallace, A. R., *The Geographical Distribution of Animals* (New York: Harper and brothers, 1876).

——*The Malay Archipelago* (New York: Macmillan, 1906).

Ward, P. D., *On Methuselah's Trail: Living Fossils and the Great Extinctions* (New York: W. H. Freeman, 1993).

Wegener, A., *The Origin of Continents and Oceans* (New York: Dover, 1966).

Wilson, E. O., *Consilience: The Unity of Knowledge* (New York: Knopf, 1998).

——*The Diversity of Life* (New York: Norton, 1999).

INDEX

INDEX

INDEX

INDEX

INDEX

INDEX

INDEX

INDEX